# DISCOVERING PHYSICS
## A Laboratory Guide

### Joel R. Hammer

*Essex County College*

**KENDALL/HUNT PUBLISHING COMPANY**
4050 Westmark Drive   Dubuque, Iowa 52002

Cover Photos: Six Flags Great Adventure Theme Park (Jackson, NJ)

*To my mother and father ,*
*Evelyn and Marty Hammer :*

*I dedicate this book , my first , to you , Mom and Dad .*
*Whatever I accomplish in the rest of my life, can be*
*directly traced to the love, sacrifices, and encouragement*
*you've given to me all of my life, and a large part of yours.*
*You shaped me, and Sheila and Marilyn (my sisters) ;*
*you taught me right from wrong ; you realized the*
*importance of education (you had to persuade me to go to*
*College, remember !) . I am the product of the loving,*
*nurturing family I grew up in. I can never give you back*
*all that you've given to me, but I'll never stop trying.*
*I love you both.*

# *To you , the EXPLORER of Physics*

*From the moment you're born, you find yourself on a ball. You find strange things moving in front of you. You see pretty colors and attractive patterns. You hear weird noises, and what you smell is sometimes nice, and sometimes stinky. You reach out and try to touch everything you see, but discover you need to do something to help you touch anything beyond the reach of your hand.*

*As a child, you stand on a chair to reach the sink your mother puts things into, and discover you can reach the handle of a closed cabinet door. When you open it, you discover a whole new world of objects, and you try to make sense of them. Some objects you know the purpose of, while others are a complete mystery to you. So you grab the unknown objects and experiment with them to see if you can discover a connection between the way they respond, and the behavior patterns you've observed when you've probed other objects, in the same way. You're acting out of curiosity and instinct. **You are a natural-born scientist !** Every human being is a natural-born scientist ; It's a quality you never lose, like learning to ride a bicycle.*

*As people evolve, in their lives, some may find they prefer to concentrate on other aspects of their life. But the tendency to want to reduce the plethora of observations to a few simple rules, is always there. How else could you explain the world-wide popularity of so many popular-science (i.e., non-technical) TV programs, magazines, books, toys, movies, etc. ?*

*The famous, much loved, and much missed physicist **Richard Feynman**, compared the "game" of discovering Physics, to a person watching a chess-game for the first time. After a while, the ways the pieces are moved, seem to conform to simple predictable patterns. Then all of a sudden, an anomaly occurs : the King jumps two boxes, instead of one (it was 'castled' !) You couldn't have predicted it, yet it happened. You realize that you now have to modify your hypotheses, to accommodate newly discovered information. As a result, your model of how the game works, changes slightly.*

*If you are lucky enough to get paid for doing this for the way nature behaves, then you are a Physicist. If you are lucky enough to be in a laboratory environment, where you're able to apply your scientific instincts, in exploring the natural world you've grown up in, then you're a college Physics student, about to embark on an exciting journey into the unknown, as you rediscover the world ! Use this Guide as your map, and treat your Professor as your knowledgeable assistant.*

----- Happy Hunting     *J. R. H.*

# To you, the <u>*TEACHER*</u> of Physics

**NOT just another lab. book**

Let's face it ; as interesting as physics is to *us*, that doesn't mean its *automatically* going to be interesting to students. This is true *even more so* in our College-Physics courses (algebra-based) as compared to our University (calculus-based) physics courses. So how should we, as consummate professionals help our students experience the joy and excitement of relating something in their everyday lives to something they've just heard about in class ?

We do this <u>in class</u> by asking thought-provoking questions, performing demonstrations, getting the students involved, relating applications to common phenomena, using humor. Some people are better at these motivating techniques than others. Think back to when *you* were taking Freshman Physics. ; Was every teacher as good as every other teacher ?

We use <u>better textbooks</u>. There has been much competition amongst textbook authors in the last few years, resulting in some excellent quality textbooks. I have reviewed many of these textbooks. One of the criteria I use for adopting a book is to ask myself the question:

> *"If I were a student reading this for the first time, would it enlighten my understanding or confuse me more than I'm already confused ?"*, i.e., *"Can I <u>relate</u> to it, or is it all just a bunch of words and equations I need for a test ?".*

<u>What about ' *hands-on* ' learning</u> ?

**Students *enjoy* doing experiments.**

Students *enjoy doing* experiments. Its a natural instinct. The more physical senses we can use, the better we can build 'mental models' to understand. *'Touch-feely'*, and *sight* are powerful tools ! Fortunately, relatively low-cost laboratory equipment has evolved over the last 30-years to the point where the majority of Colleges and Universities have Air-Tracks, Photogate-Timers, and Lasers, tools that eliminate unwanted phenomena and improve precision so the student can concentrate on the relationships between the important physical variables. Half of the experiments utilize this equipment.

<u>**Laboratory manuals**</u> are the <u>***weakest link***</u> in the physics education of our students !

⇉ Existing Laboratory experiment books are <u>BORING</u> , <u>DRY</u> , <u>DUSTY</u> , and <u>MECHANICAL</u> !! Invariably, they seem to be written more as an anthology of robotic procedures, than as an interface to human-beings ! Its no wonder how easily a student can get ' turned-off ' when being forced to carry them out without being 'psychologically motivated' .

*" Putting the Cart Before the Horse "*

⇉ Another <u>*major problem*</u> all existing lab manuals share, is that they invariably expect the student to understand the physics <u>*before*</u> the student does the experiment, by presenting the relevant equations first ! This a classic example of putting the cart before the horse. If anything could kill enthusiasm, it would be thinking that what it all boiled down to

was putting some numbers in an equation that was literally 'pulled out of thin air' !

saves the student $$$$ & conserves trees

⇉ Another fact of life is that few professors, if any, have their classes do more than 15–20 experiments in an academic year. So why should students have to waste their money buying lab books of 60 experiments when they wind up doing less than 1/3 of them ? This book contains 17 important experiments commonly required of students to perform in a one-year College Physics sequence. Each of the experiments can be performed in one class session of 1:$^{Hr}$ 20$^{Min}$ .

I wrote this book as a physics laboratory *guide* to help *students* <u>*DISCOVER*</u> *physics*. I wrote it for *them* , not for me. In it, I use some of the same techniques that I find successful in the classroom : humor, relating everyday phenomena, suspense, anticipation, and thought-provoking questions. I talk to them as what they are: intelligent human-beings who have skills but not knowledge of a particular topic. My approach seems to work well ! This material has been used now for several years by many students and professors ; The positive reaction I have received has prompted me to offer it to a wider audience. I wish you the same success.

I want to thank my colleague, Professor August Ruggiero, at Essex County College, for voluntarily taking on the burden of smoothing out the many wrinkles in my original manuscript, as it evolved over time. I'm grateful to him for spotting and correcting not only grammatical mistakes, but also inconsistencies in some of the experimental procedures.

----- Enjoy the experience !        *Joel R. Hammer*

# COMING ATTRACTIONS

# MEASURING the WORLD

How old are you ? How tall are you ? How fast can you run ? What's the temperature ? The answers to these questions becomes meaningless without associating the numbers with a physical unit, like years, feet, meters per second, or degrees Fahrenheit.

But measuring these physical quantities directly, implies an understanding of the effect the measuring instruments you use, have on the way you state your answer. For example, measuring your height with a ruler, then again with a LASER and a Mirror, can't possibly give the same answer ! (This is the method that NASA used to determine the distance of the Moon from the Earth (approximately 240,000 miles), by connecting the LASER to a computer to time the travel time of the beam after reflecting off a mirror placed on the Moon's surface by Astronauts. The answer was precise to within *ONE FOOT !!!* )

The inherent *uncertainty* of the measuring process, can be expressed in several different ways, that we'll discuss. All *errors* that can be associated with the act of measurement, fall into three categories: human, systematic, and random. The better you understand the physical conditions under which you perform your experiments, and the nature of the instruments you use, the more believable your results will be.

# The TOOL-BOX:

| | | |
|---|---|---|
| Meter Stick | You | various objects: Cylinder, Sphere |
| Ruler | Stop-Clock | Mass Balance |
| Vernier Caliper | Photogate Timer | Air-Track system (Blower, Hose) |
| Physics Textbook | | |

# Part A: *ARE YOU Average ?*

"The *average* American family has 2 1/2 children". This government statistic begs the question: Exactly which parts of a half-child is missing ? It is also an excellent illustration of an important statistical fact:

**THE AVERAGE OF A SET OF NUMBERS
DOESN'T HAVE TO BE THE SAME AS
ANY OF THE NUMBERS IT REPRESENTS !**

In the Physics laboratory, a single measurement can be misleading. Perhaps one of your partners bumped into the lab table; perhaps a gust of wind came by. Whatever the reasons, the careful experimenter makes several TRIAL RUNS, and then calculates the mathematical average of the set to use as a representative value. Poll your group to determine its average age, by adding the ages, and dividing by the number of group members.

For example, if there are four group members:

$$x_{ave} = \frac{x_1 + x_2 + x_3 + x_4}{4}$$

# Part B: *HOW* **Significant** *DO YOU* **Figure** *IT IS ?*

If you were measuring the length of the above bar, with the centi-meter ruler, above, what would you say it is ?  3 cm. ? ,  4 cm. ? , 3.725 cm. ?  A scientist would say "NONE of these choices are very good; 3.7 cm. is more realistic, or even  3.5 cm. ! "

Certainly the bar is much greater than  3 cm. , but somewhat less than 4 cm.. It's not difficult to mentally break up the smallest interval,  one-cm., into two pieces, or even into ten pieces, with a little practice.  It wouldn't be unreasonable, then, to estimate the length to the *nearest half-centimeter, or even the nearest tenth of a centimeter.* Your estimate might be slightly different from your partner's estimate, but what's important is that your answer doesn't convey *too much* or *too little* information about the instrument you used.

The situation becomes even worse when you have to perform mathematical operations, such as multiplication, division, addition, and subtraction, on numbers obtained from direct measurements made with different measuring instruments !

To avoid chaos, scientists have agreed upon a useful set of rules for describing and manipulating quantities.  The topic is called **SIGNIFICANT FIGURES** .

---

➠ **RULE** ①     The number of significant figures  =  the number of digits you can read
EXACTLY , from the instrument, **PLUS ONE** for the **ESTIMATE** of
the **FRACTION OF THE SMALLEST DIVISION.**

---

➠ **RULE** ②     The digits  **1, 2, 3, 4, 5, 6, 7, 8, 9**  are  **ALWAYS SIGNIFICANT.**

---

➠ **RULE** ③     A **ZERO**, sitting **BETWEEN** significant digits,  is  SIGNIFICANT.
ex:  2001 (four)            1020309 (seven)

---

➠ **RULE** ④     **ZEROS** on the *RIGHT END* of *WHOLE NUMBERS*, are  **NOT SIGNIFICANT**,
*UNLESS MORE INFORMATION* about the measuring instrument is given.
These zeros simply tell us what power of ten we are dealing with, but not how
precise the instrument is. An example with  2  significant figures is:
Sun-to-Earth- distance  =  93,000,000 miles

---

➠ **RULE** ⑤     **ZEROS** to the **IMMEDIATE RIGHT** of **DECIMAL POINTS** are:
a)  SIGNIFICANT if the number is ONE or LARGER
ex: 43.0 (three)         191.04 (five)         1.000000000 (ten)
b)  NOT SIGNIFICANT if the number is LESS THAN ONE
ex: 0.0032 (two)              0.015 (two)

---

---

⮕ **RULE ⑥**   **ZEROS** on the **FAR RIGHT** of **DECIMAL POINTS** are  **ALWAYS  SIGNIFICANT**.

           ex:  4.3000  (five)          0.00500  (three)

           ex:  4.30000  (six)         1.00500  (six)

---

⮕ **RULE ⑦**   When numbers are written in  **SCIENTIFIC NOTATION,  ALL DIGITS ARE  SIGNIFICANT !**

           ex:  $3 \times 10^8$  (one)       $3.00 \times 10^8$    (three)

===

# Part  C:  THE BIG *Round-Off !*

     You might refer to that **$6.98** flashlight you bought, as a **$7** flashlight.  If you did its because you discovered that it's sometimes convenient to round numbers to fewer places.  Numbers, obtained from *physical measurements*, when rounded off properly, convey information about the measuring instrument ; namely its *precision*.  For example, measuring the width of your living room with a tape measure is more precise than counting your footsteps placed end-to-end.

     We characterize the precision of an experimental measurement we make, using a particular instrument, by the number of significant figures.  So, when we round-off a number to the appropriate number of significant figures, we usually wind up having to drop some digits.

    The Rules-of-Thumb commonly used for rounding-off are :

---

(Always *start* by examining the *1st digit to the  right* of the *last significant digit*.)

Ex:    If the number has  7  digits:    ☐☐☐.☐☐☐☐ ,

and we want to round it off to  4  sig. fig.:    ☑☑☑.☑◯☒☒ ,

then we start with this digit:    ———————→ ↗

---

❶  If it is  **0,  1,  2,  3**, or  **4**, then drop it and all the digits after it.

     Ex:  Rounding off to  3  significant figures:   8.2341

     Answer:   8.23    (the 4th digit is less than  5)

❷  If it is  **6,  7,  8**, or  **9**, then

     a)  drop it and all the digits after it,  *and*

     b)  add one  to the last significant digit you kept.

          Ex:  Rounding off to  3  significant figures:   8.2372

          Answer:   8.24    (the 4th digit is greater than  5)

❸  If it is  **5**, and the digits which follow it are **not** zero , i.e., 1,2,3,4,5,6,7,8,9

     then follow the same two steps in  ❷ .

          Ex:  Rounding off to  3  significant figures:   8.2351234

          Answer:   8.24    (the 4th digit is 5 , and the 6th, 7th, and 8th digits are non-zero)

❹ If it is **5**, <u>and</u>

     ↳ it is the *last digit*,     Ex:  8.23**5**

<u>or</u>  ↳ the digits which follow it are *all zeros*,  Ex:  8.23**5**00

then drop it and all the digits after it, and then use the *ODD-EVEN rule* :

---

### ODD-EVEN rule

a) If the last significant digit is <u>EVEN</u> ,
        leave it alone (you're finished !).
b) If the last significant digit is <u>ODD</u>,
        <u>add one</u> to that last significant digit .

---

Ex:  8.23**5**  =====➤  8.24     Ex:  57.6**5**  =====➤  57.6
Ex:  8.23**5**00 =====➤  8.24     Ex:  57.6**5**00 =====➤  57.6

◉ ◉    ⎛ *To help you see the rules at a glance,*
                *check out the flow chart*
                      *on the next page !* ⎞

---

**Q**uestion:   " But what if I have to ***put the data into a formula***, and use my calculator to
                 figure out the answer ?   How many significant figures do I keep ? "

**A**nswer:   *Good question !  I'm glad you asked !*    Here's an idea that will give you a clue:
             If you had a chain all of whose links, except one, are made out of heavy-duty
steel, and the one different link is made out of rubber, and you  p--u--l--l--e--d
the ends of the chain in opposite directions ⇐ ⇒ , *where would it snap ?*
            It would snap at the *weakest link*, the rubberband, i.e., *a chain is no stronger
than its weakest link !*
       Similarly, *the answer from a calculation can't be more precise than the least
precise measurement that went into the calculation*.

So,   
> ROUND OFF YOUR ANSWER TO THE SAME NUMBER OF SIGNIFICANT
> FIGURES AS THE NUMBER WITH THE <u>FEWEST NUMBER OF
> SIGNIFICANT FIGURES</u> !

Ex:    $\dfrac{6.2749}{1.44}$  =  4.357569444  (on calculator)        Correct Ans:  4.36

# THE BIG rOUND–OFF Flow Chart

Always *start* by examining the *1st digit to the right*
of the *last significant digit*.

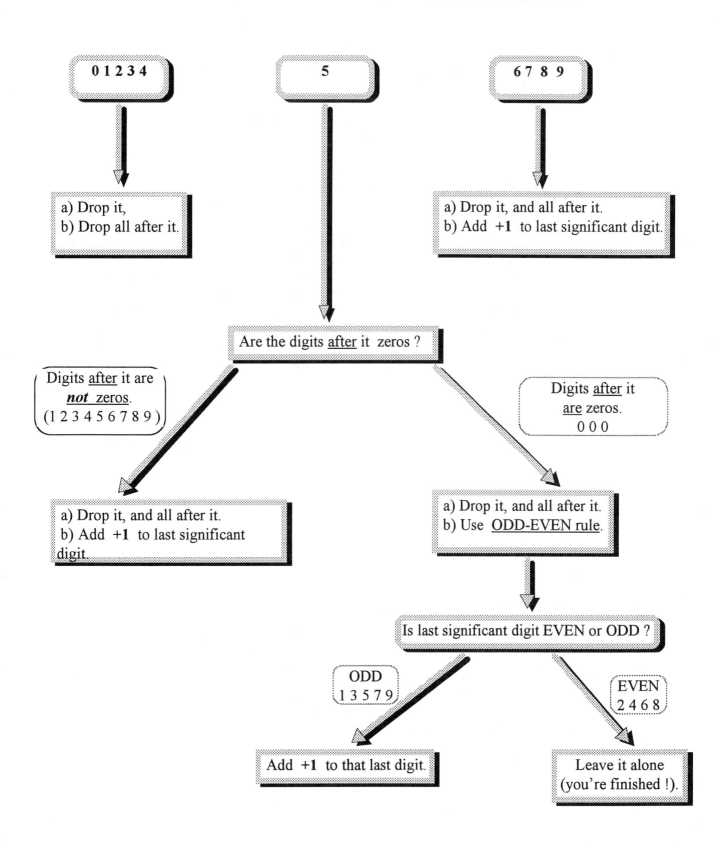

# Part D: *Errors are* NOT *just math mistakes !*

There are THREE TYPES of MEASUREMENT ERRORS:

1) <u>**GOOF-UPS**</u> - Anyone can make a mistake.  Minimize its occurrence by having more than one group member repeat the *observations* and *calculations*, then *check each other*.

2) <u>**SYSTEMATIC ERRORS**</u> - If the instrument is not aligned to the correct reference point, all the readings will be "off" **(shifted) by the same amount**.  This error is sometimes difficult to spot, but easy to fix !

   ex:  not putting an object at the exact zero position of a ruler

   ex:  You come two minutes late to *all* of your classes, every day, because you carelessly set your wristwatch two minutes behind the last time you adjusted it.

3) <u>**RANDOM ERRORS**</u> - No matter how carefully produced, "identical  objects" cannot be perfectly alike. Tiny differences in shape, mass, and molecular structure will always exist.  You cannot even reproduce the same environmental conditions of a measurement.  The difference becomes noticeable when you repeat a measurement several times, like tossing a penny on the table.  Try it now !  Toss a coin  10 times and count the number of HEADS.  Let each of your partners do the same thing with the *same coin !*  Now compare results with each other.  If you didn't get  5 heads, it's not because you did something wrong, it's because tiny random differences in the physical conditions of the toss, affected each outcome.  Each person's results may not be the same, for the same reason.  The physics of random phenomena can be explored with a knowledge of statistics, but for now, two ideas are important:

   ☞     Learn to recognize *when* you are dealing with random phenomena, and, if possible, what the specific random variables are.

   ☞     Try to repeat a measurement several times (the more trials, the better), and then calculate the average value.

---

# Part  E:  *'Got  a  Minute' ?*

<u>**Goal:**</u>   To discover your accuracy and precision as a "human stop-watch".

<u>**Theory:**</u>  The human brain and the five senses can be relied upon to do many things, but sensing the passage of time, accurately, is not one of them !  You can test this assertion on your own group by using the stop-clock to time how long it takes each of your team members to perceive the passage of one minute, and calculating the error.

## Procedure:

1) Synchronize the start, by having the "human-clock" begin counting the passage of what he/she believes is **60 seconds - one minute**, at a  pre-arranged  signal. Make sure the stop-clock is not visible to that member.    (No cheating !)

2) The "human-clock" should count,  out-loud  during the last  five  or so    "perceived" seconds, so that the person controlling the electrical stop-clock will  be warned when the end is coming.

3) When the "human-clock" says " SIXTY ",  the electrical stop-clock is stopped, and this experimental time is recorded as  $t_{exp}$ .

4) Each member of the group should get a turn at testing themselves.

## Calculation:

1) Calculate your **time difference**, with respect to  60 seconds :

$$\text{time difference} = \Delta t = t_{exp} - 60$$

2) Calculate the **Per-cent Error** of the times, with respect to  60 :

$$\% \text{ Error} = \frac{\Delta t}{60} \, 100\,\% = \frac{t_{exp} - 60}{60} \, 100\,\%$$

---

# Part  F:  *HOW   Thick  IS  YOUR   Page ?*

**Goal:**        To **measure the thickness** of a  *SINGLE PAGE  of your*  TEXTBOOK !

**Theory:** Each *physical* page in your Physics textbook has printing on each of its two sides.  Let's first consider the sides numbered from  1  to  800 .  The number of corresponding *pages* is, therefore, **half** of the amount of numbered sides.

number of pages  =  1/2 (800)  =  400 pages

If you measure the thickness of those 400  pages, by placing the small ruler next to them, you can calculate the average thickness of a *single page* by dividing the your measured thickness by the  400 pages:

$$\text{thickness of a } single\ page = \frac{\text{ruler measurement}}{\text{number of pages}}$$

## Step-by-Step :

(1)  Open your Physics textbook, and stand it up on the table, facing you.

(2)  Hold only the pages whose sides are numbered  1  through  800 , with one hand,
       so that this hand passes over the top of the book.  Don't squeeze the pages
       too hard !

(3)  Pick up the short ruler with your other hand, and place it against the flat edge
       of the pages at the top (the same edge you are holding).

(4)  Measure the thickness of these pages directly with the ruler, to the proper
       number of significant figures, and enter it in your data sheet under *Trial 1*.

(5)  Do *two more*  Trials, using *different* group members, for the same # of pages.

(6)  Repeat steps  #2,  #3,  #4 , and  #5 , but for each of the following ranges of
       pages numbered:      1–500 ,  1–300 ,  1–100 ,  and  1–20 .

## Part  G:   THE *Vernier - Caliper* , ANOTHER TYPE OF "RULER"

A more precise, yet still low-cost, "ruler" is the Vernier-Caliper.  Even the simplest Vernier measures to *1/100 cm* , *with certainty !*  Adjust the sliding Vernier scale to accommodate the object in the Calipers (the sliding Vernier is rigidly attached to the Calipers).

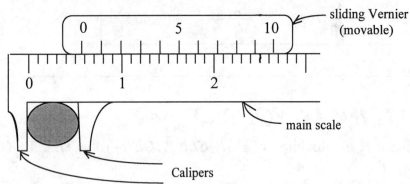

The measurement is the *sum* of two numbers: (1)  the largest *main-scale* number before the zero of the Vernier (in the diagram it would be 0.5 ), and (2)  the Vernier number that EXACTLY MATCHES ANY main-scale number (in the diagram it would be 3 ).  (Since the Vernier has 10 intervals, it, in effect, divides the smallest main-scale division into  10 parts, thereby giving us that extra significant figure !)

The width of the object in the diagram is:

$$0.5 \text{ cm} \quad + \quad (\textit{three} / 10) \ (0.1 \text{ cm.}) \quad = \quad 0.53 \text{ cm.}$$

## Part  H:   *It's the REAL Thing !*

Let's put all of your newly acquired skills to the test, by determining the density of some objects.

➜ ALWAYS KEEP IN MIND, THAT

THE NUMBER *YOU RECORD* IN YOUR DATA,
BASED ON A DIRECT MEASUREMENT,
MUST HAVE THE
*CORRECT  NUMBER  OF  SIGNIFICANT FIGURES*
THAT THE *MEASURING INSTRUMENT ALLOWS !!!*

## Step-by-Step :

1) Measure the <u>mass</u> of the cylinder, and of the sphere, on the mass balance.
2) Using the Vernier-Caliper, measure the <u>diameter</u> of the cross-section of the cylinder, as well as the <u>diameter</u> of the sphere, and record both in your data sheet.
3) Using your small ruler, measure the <u>length</u> of the cylinder.
4) Repeat steps  #1 - #3  for two or three more trials.  (The best way to do this is to let each group member perform a complete trial set of measurements.)

*All my trials*

## Calculations:

1) Calculate the volumes from the following formulas:

VOLUME of a SPHERE   =   $4/3 \, \pi \, r^3$   =   $1/6 \, \pi \, (\text{diam})^3$

VOLUME of a CYLINDER   =   $(\pi r^2) \, \ell$

= $1/4 \, \pi \, (\text{diam})^2 \, (\text{length})$

2) Calculate the  DENSITY of both objects, using the definition of density:

$$\text{Density} \quad = \quad \frac{\text{mass}}{\text{Volume}}$$

# Part I:  *Gliding  On  Air* , IN THE *Blink of an  Eye !*

Two additional high-precision measuring instruments that you'll be using  frequently throughout the semester are:

➪   a)  the  AIR-TRACK SYSTEM, and

➪   b)  the  PHOTOGATE-TIMERS

Their **purposes** are to:
1) help extend the range of your senses
2) minimize influences that can interfere with the fundamental
   phenomena being observed, such as *friction*, and *human nervous
   system reaction time*
3) familiarize you with their construction, operation, and maintenance.

After your instructor has described them to you, and if there is enough time
remaining today, use them to make a measurement of a time-interval.

**Remember:**

> *What goes around, comes around.*
>
> *i .e.,*
>
> *If you are good to your instruments,*
> *your instruments will be good to you !*

# DATA  &  CALCULATIONS

Wherever a set of parentheses is shown, fill in the correct unit of the corresponding measured
and computed physical quantities.

### Part A :   *Are You AVERAGE* ?:

$(Age)_1$ = _____        $(Age)_3$ = _____

$(Age)_2$ = _____        $(Age)_4$ = _____

$(Age)_{ave}$ = _____        Show
your work

### Part D3 :   *Random-ERRORS* - Same coin, tossed 10 times, randomly:"

number of  HEADS    = _____

### Part E :   '*Got  a  Minute* ?'

$t_{exp}$ = _____

$\Delta t$ = _____        % error = _____ %

## Part F :  *How  THICK  is  your  PAGE ?*

| No. of pages (N) | Trial #1 | Trial #2 | Trial #3 | Ave. Total Thickness (T) | Calculated thickness of one page (t) |
|---|---|---|---|---|---|
| 800 | | | | | |
| 500 | | | | | |
| 300 | | | | | |
| 100 | | | | | |
| 20 | | | | | |

(Ave)  single page  thickness        $t_{ave}$   =   _____

## Part H :  *Its the REAL thing !  -  Density of Sphere  and  Cylinder*

SPHERE

| Trial No. | Diameter (cm.) | Mass (gm.) |
|---|---|---|
| 1 | | |
| 2 | | |
| 3 | | |
| 4 | | |

CYLINDER

| Diameter (cm.) | Length (cm.) | Mass (gm.) |
|---|---|---|
| | | |
| | | |
| | | |
| | | |

(Ave) Density of <u>Sphere</u>  =  _____          (Ave) Density of <u>Cylinder</u>  =  _____

# QUESTIONS :

1.  Is anyone in your group "average", using the average age ?

_____

2.   When an average must be calculated from a set of trial runs, but ONE value is *much different* than all
     the others, would you include it or ignore it ?   Why  or  why not ?

     _____

     _____

     _____

     _____


3.   What's wrong with using a Vernier Caliper directly on a single page ?

     _____

     _____

     _____

     _____


4.   How would you measure the  VOLUME  of an  IRREGULARLY  SHAPED  object ?

     _____

     _____

     _____

     _____


5.   In Part F , does the accuracy of " t " increase with an increase in  N ?   Why or why not ?

     _____

     _____

     _____

     _____

## AVERAGE VELOCITY and INSTANTANEOUS VELOCITY

The distinction between *average velocity* and *instantaneous velocity* can be a sensitive thing to someone. Have you ever seen a runner practicing doing intervals around an oval track? Perhaps it's you ! Well, after doing a hard workout of several laps, *don't ask that runner what his/her ' average velocity'* was ! The answer is *zero !* The *average velocity* is (the final position – the original position) per unit time. For any number of whole laps, the positions are the same ! A more appreciative question might be ' What was your *average* speed? ' , or ' What was your fastest *instantaneous* speed ? ' Even if we were talking about motion in a straight line, we have to distinguish between *average velocity* and *instantaneous velocity*. It's relatively easy to measure an *average* velocity , but how can you measure the velocity at a point ? In this experiment you'll investigate the relationship between instantaneous and average velocities, and see how a series of average velocities can be used to deduce an instantaneous velocity.

# The TOOL-BOX:

|  |  |
|---|---|
| Air-Track | Photogate Timer |
| Air Supply | Accessory Photogate |
| Air Hose | Small Gold Glider |

100  Displacement

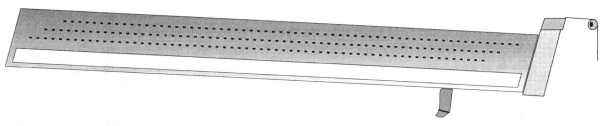

Figure 1

# STEP-BY-STEP:

1. Set up the air track as shown in Figure 1, elevating one end of the track with a 1–2 cm. support.  Use a rigid object for the support, not a "mushy" one.

2. Choose a point near the center of the track, that we'll call $x_1$ .  Observe the position of this point on the air track's metric scale, and record this value on your data table.

3. Choose a starting point $x_0$ for the glider, near the upper end of the track.  Record this position in your data table.  This is the position that you'll always start your glider from.

4. Place the Photogate Timer and the Accessory Photogate at points equidistant from $x_1$ (see " On your Mark ! " in sidebar),  as shown in Figure 1. Record the distance  **D** between the photogates, in your data table.

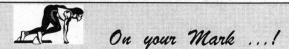

*On your Mark ...!*

Since its difficult to judge, by eye, *precisely* where the Photogate beam is, here's a *very precise* way of doing it:

- Put the front of the glider exactly at the location you want the Photogate beam to be at.
- *V-e-r-y s-l-o-w-l-y* guide the Photogate toward where the glider is.
- When the red light on top of the Photogate goes ON, *STOP moving the photogate*.  The beam is precisely at that position of the front of the glider..

5. Set the slide switch on the Photogate Timer to  PULSE (Fig. 2).

6. Press the  RESET button.

7. Hold the glider steady at $x_0$ , then release it.  Record time $t_1$ , the time displayed after the glider has passed through both photogates.

8. Repeat steps  6  and  7  four more times, recording each of the times as  $t_1$  through $t_5$ .

9. Now repeat steps  4  through  9 , decreasing  **D**  by  10 centimeters.

10. Continue decreasing  **D**  by  10 centimeters. At each value of  **D** , repeat steps  4  through  8 .

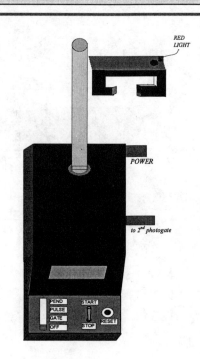

Figure  2

# Show-the-World:

1.  For each value of **D** , calculate the average of times $t_1$  through $t_5$ :
    $t_{avg}$   = Record each average time in the  $t_{avg}$  column of your data table.

$$t_{avg} \quad = \quad \frac{t_1 \ + \ t_2 \ + \ t_3 \ + \ t_4 \ + \ t_5}{5}$$

2.  For each value of **D** , calculate the average velocity of the glider in going between the two
    photogates:

$$v_{avg} \quad = \quad \frac{D}{t_{avg}}$$

    and record in the  $v_{avg}$  column

3.  Plot a graph of  $v_{avg}$   versus  **D** ,  with **D** on the horizontal axis.

# DATA:          *Measuring Mother Nature*

$x_1$ = _____          $x_0$ = _____

| D (cm) | $t_1$ (sec) | $t_2$ (sec) | $t_3$ (sec) | $t_4$ (sec) | $t_5$ (sec) | $t_{avg}$ (sec) | $v_{avg}$ (cm/sec) |
|---|---|---|---|---|---|---|---|
|  |  |  |  |  |  |  |  |
|  |  |  |  |  |  |  |  |
|  |  |  |  |  |  |  |  |
|  |  |  |  |  |  |  |  |
|  |  |  |  |  |  |  |  |
|  |  |  |  |  |  |  |  |
|  |  |  |  |  |  |  |  |
|  |  |  |  |  |  |  |  |
|  |  |  |  |  |  |  |  |

# QUESTIONS:　　　*Did you Learn Something ?*

1.  Which of the average velocities that you measured do you think gives the closest approximation to the instantaneous velocity of the glider as it passed through point $x_1$ ?

2.  Can you extrapolate from your data to determine an even closer approximation to the instantaneous velocity of the glider through point $x_1$ ? From your collected data, estimate the maximum error you expect in your estimated value.

3.  In trying to determine an instantaneous velocity, what factors, for example: timer accuracy, object being timed, type of motion, etc.) influence the accuracy of the measurement ? Discuss how each factor influences the result.

4.  Can you think of one or more ways to measure instantaneous velocity directly, or is an instantaneous velocity always a value that must be inferred from average velocity measurements ?

What does this familiar description of a new car:

*"It goes from zero to 60 in 8 seconds"*,

have in common with:

*"the Moon revolves around the Earth"* ?

Both describe a *change* in velocity, i.e., an ACCELERATION. The car's data is based on a <u>change in speed</u> along a <u>straight</u> track ; The Moon's changing velocity is due to it's continually <u>changing direction</u> (a topic you'll explore in Exp't 8 ).

What you're about to investigate, both experimentally and graphically, is the specific relationship between instantaneous velocity and elapsed time along a straight hill. These "investigative tools" will give you a deeper insight into the meaning of uniform acceleration.

# The TOOL-BOX:

Air-Track       Photogate Timer
Air Supply      Small Gold Glider
Air Hose

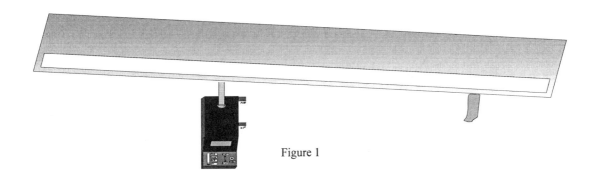

Figure 1

# STEP-BY-STEP:

1. Set up the equipment as shown in the diagram. Place some textbooks or other rigid objects underneath the support leg at one end of the air track. Their combined thickness should be about 2–3 cm.

2. Set the Photogate Timer to GATE mode and press the RESET button.

*On your Mark ....!*

3.  Make a precise measurement of the length of the glider,  **L** ,  by doing the following: While holding the glider,  *v-e-r-y s-l-o-w-l-y* guide it through the Photogate, and record the scale position,  $L_0$  , at which the LED light on the Photogate Timer first goes on.  Continue guiding the glider until the LED light goes off, and record the location of the front of the glider as  $L_1$ .  The difference of these two values is the effective length of the glider.  Now, divide  **L**  by  2 ,  and record this value as  $\Delta L$ .  This represents the <u>midpoint</u> of the glider.  Add  $\Delta L$  to  $L_0$ , recording this as  **x**:

$$x \ = \ L_0 \ + \ \Delta L \ .$$

4.  Press the  RESET  button on the Photogate Timer.

5.  Place the glider on the airtrack so that it's front end is  10 cm. away from  **x**  on the higher side.  You can use a ruler to hold it in place, while your partner operates the Photogate Timer.  Now quickly remove the ruler, allowing the glider to slide down the hill and move completely through the Photogate.  Record this time as  $t_1$ .

<u>*All My Trials*</u>

6.  Repeat steps  4  and  5  two more times, so that you will have a total of three trial runs for  10 cm., and record all three times in your data table.

7.  Let's now get a set of three trial runs, at distance of  20 cm., instead of  10 cm.  Repeat steps  4  to  6 , but place the glider  20 cm.  away form the midpoint location.

8.  Repeat steps  4  to  6 , for distances of  40 cm.,  60 cm.,  80 cm.,  120 cm., 140 cm.,  and 160 cm., respectively.

# Show-the-World:

1.  Calculate the average time of the set of three trial runs that correspond to each value of  **x** ,  by adding them and dividing by three:

$$t_{avg} \ = \ \frac{t_1 \ + \ t_2 \ + \ t_3}{3}$$

Record  t  in your data table.

2.  You can now calculate the final velocity of the glider for each distance  **x** ,  by dividing  **L**  by the corresponding average time.  Enter this as  **v**  in your data table.

3.  You can get a better insight into the pattern of the motion of the glider by constructing some graphs of your data.

    a)  To analyze how fast the glider is moving after having traveled various distances, you need to plot a **velocity vs. distance** graph, with the distance  **x**  as your horizontal axis.  If the graph doesn't turn out to approximate a straight line (it shouldn't), then let's manipulate the data to see what it takes to obtain a linear graph.

b) For each   distance calculate $v^2$ , and   $1/v$   and enter these into your data chart.

c) *Plot **three** graphs*, with   $x$   on the horizontal axis of each, and   $v$ ,   $v^2$ ,   or   $1/v$   on the vertical axis, respectively

d) Plot a **fourth** *graph* of   **velocity   vs.   time** , with **time** on the horizontal axis.   It should be a straight line.

4.   Calculate the *slope* of your **velocity   vs.   time** graph, and record it on your data sheet.

5.   Calculate the *slope* of your   $v^2$   **vs.   x**   graph , and record it on your data sheet.

# DATA:   *Measuring Mother Nature*

$L_0$ = _____ cm          $L$ = $L_1 - L_0$ = _____ cm

$L_1$ = _____ cm          $\Delta L$ = $L/2$ = _____ cm

$x$ = $L_0 + \Delta L$ = _____ cm

| x ( cm ) | $t_1$ (s) | $t_2$ (s) | $t_3$ (s) | $t_{avg}$ (s) | | v (m/s) | $v^2$ $(m/s)^2$ | $1/v$ (s/m) |
|---|---|---|---|---|---|---|---|---|
| | | | | | | | | |
| | | | | | | | | |
| | | | | | | | | |
| | | | | | | | | |
| | | | | | | | | |
| | | | | | | | | |
| | | | | | | | | |
| | | | | | | | | |

slope of   **v**   vs.   **t**     graph   =   _____

slope of   $v^2$   vs.   **x**     graph   =   _____

# QUESTIONS:        *Did you Learn Something ?*

1.  Which of your four graphs most closely approximates a straight line ?  This is one way a scientist, or any curious person (hopefully you!) would approach an understanding of an unknown phenomena.

2.  To gain further insight into why the relationship isn't linear, let's examine the equations of uniform motion that have time in them, namely: $x = 1/2\, a\, t^2$   and   $v = a\, t$ . Solve the $2^{nd}$ equation for $t$ , and substitute it into the first equation.  You now can see that it's the *square* of the velocity that is directly related to the distance.

3.  But what's the physical explanation of what has happened?  Is the distance traveled directly proportional to the time elapsed?  If the fastest speed you could run is twice as fast as your friends fastest speed, would it take you half as much time to race him/her over the same distance, from rest ?  Discuss your answers and your reasoning.

4.  Using dimensional analysis, determine the physical units of the **slope** of your **velocity vs. time** graph.  Do you think that your slope represents the acceleration of the glider down the airtrack, (within a small experimental error)? If so, why ? If not, why not ?

5.  Is the slope of your $v^2$ **vs. x** graph the same (within a small experimental error) as the slope of your **velocity vs. time** graph ?  If not, what would you estimate their relationship to be ?  What do the equations of motion have to say about it ?

# PROJECTILE MOTION

## 4

A high jumper is concentrating on jumping (or pole-vaulting) as high as possible, without landing too far away. A broad-jumper like Carl Lewis, tries to land as far away as possible, without jumping too high. To a golfer, a shot-putter, an Air-Force bomber pilot, a baseball player, an Astronaut on the Moon, and a parachutist, predicting *exactly where* they (or the projectiles they release) are going to land, could mean the difference in whether they *live or die*, or at the least, have an influence in their career.

Every one of these people understand the laws of Physics that Nature imposes on freely falling objects, in two or more dimensions.

The graceful curve that water droplets, from a fountain, follow, is based on the idea that the vertical acceleration of gravity has <u>no influence</u> on where the landing point is !

You're about to explore, experimentally, what determines the shape of that curve.

# The TOOL-BOX:

| | | |
|---|---|---|
| Packard's Apparatus | Thin Carbon paper | Ruler |
| Metal Ball | 10 mm x 10 mm Graph Paper | Masking Tape |

# PLANNING AHEAd

1. It takes very little time to fall a short distance (the acceleration of gravity is relatively *large* ). To make it easier to measure the trajectory, we'll use Galileo's trick of *"slowing"* the acceleration, by allowing the metal ball to roll on a hill, as shown in Figure (1).

2. To give the ball a starting velocity, horizontally, we glued a small ramp to the main board, and placed a small clamp on it so that you can lean the ball against it and thereby be guaranteed a repeatable starting place.

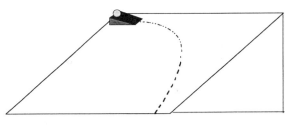

Figure 1

3. With a sandwich of carbon paper on graph paper placed on the main board, the ball will roll down the ramp, and then the main board, where it makes a beautiful trace of its trajectory !

4.  By placing the origin of an x - y coordinate system at where the ball first hits the main board, you can mark off equal horizontal displacements, allowing you to measure the corresponding vertical displacements.  You now know *exactly where* (within experimental error) the ball was at *every* instant of its motion !

# STEP-BY-STEP:

**PRACTICE**

1.  Place a solid object underneath the main board, near the top, and a *heavy* solid object, like your textbook, at the bottom, but *not underneath* (this will prevent the board from slipping).

2.  Place the graph paper, face-up, on the board, and have someone hold it.  Don't tape it yet !  Align the graph paper horizontally, and slide it a centimeter or two under the end of the ramp.

**MAKES**

3.  Place the ball against the clamp, and let it go.  To make your later analysis easier, adjust the graph paper slightly, so that the point of first contact coincides with the upper-left ruled corner of the graph-paper (make sure it's still horizontal !).  Now tape the margin with a small piece of masking tape.

4.  We're trying to get the ball roll over as much of the paper as we can, because "spreading out" the curve will make it easier for you to make more accurate measurements.  Adjust the "steepness" of the hill by sliding the object underneath it in closer, or out farther, and test it out by allowing the ball to roll from the starting position.

**PERFECT !**

5.  *Very gently*, place the thin carbon paper over the graph paper, carbon-side down, and *gently* tape it with a small piece of masking tape. (Make sure nothing moves !)

●↘  *LET 'EM ROLL !*

6.  Place the ball at the start, and let it go;  Mother Nature will then draw, for you, a smoother curve than any machine ever could !  To make the trace darker, let the ball roll a second time, *before* removing the tapes, and the papers.

✌  *TWO FOR THE PRICE OF ONE !*

7.  *HOLD IT !*  Don't remove the tapes and papers just yet !  Let's do one more experiment.  Let's investigate how the steepness of the hill affects the shape of the curve.  Slide the support underneath the board a small distance in, or out, to change the slope of the hill, and repeat step 6 .

## FINALLY !

8.  *Carefully* remove the pieces of tape.  Label your axes,  **x**  and  **y** , as shown in Figure (2).
    Since your graph paper is metric (10 mm. by 10 mm.), scale your axes by writing the
    integers  0, 1, 2, 3, ...  at every  *centimeter*  line (i.e., every  10 mm. ),  for both the  **x**
    and  **y**  axes.  Label your higher curve  " A "  and your lower curve  " B " .

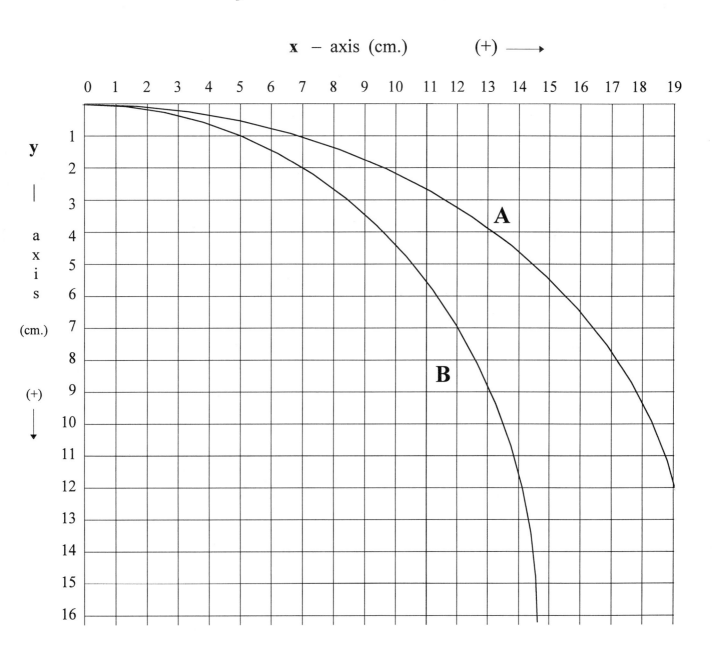

*Figure  2*

# Show-the-World:

1.  For each ONE-CENTIMETER integer on your **x**- axis, follow the vertical line down
    until it hits your " A " curve. *Neatly* mark it with a small dot (called the "data
    point") and a small *neat* circle.

2.  Carefully line up your ruler so that it passes through your "data point" , while still
    being parallel to the nearest horizontal line.

3.  To the precision of the graph paper, record the value on your **y** - axis where the ruler
    intersects your vertical scale, in your Data chart.

4.  Repeat steps 1 through 3 , for curve " B "

5.  At the point *on* curve " A ", whose **x** coordinate is 14.0 cm. , place your ruler
    *tangent* to the curve, and draw a straight line about 6 cm. long (about 3 cm. on
    each side).

    > Q:  *How can I tell it's tangent  ??*
    > A:  You can estimate it by looking at the  " white space " between the
    >      ruler and the curve, on both sides, and then adjust the angle of
    >      the ruler until they seem approximately equal.

6.  Choose two "convenient" points *on* your straight line to enable you to measure it's
    slope.  Dot them and circle them.

    > Q:  *What would make the points convenient  ??*
    > A:  Choose points that are as close as possible to the corner of any
    >      square millimeter box.  Your chances of finding ones that intersect
    >      are good !

7.  Locate the **x** and **y** coordinates of the two points by using the ruler method, and
    record the four positions in your Data sheet.

8.  Calculate the slope of the line:

$$ \text{slope} \; = \; \frac{\text{rise}}{\text{run}} \quad = \quad \frac{\Delta y}{\Delta x} \quad = \quad \frac{y_2 \; - \; y_1}{x_2 \; - \; x_1} $$

# DATA:     *Measuring Mother Nature*

### Trajectory **A**

| DATA POINT | x [cm.] | y [cm.] |
|:---:|:---:|:---:|
| 1 | | |
| 2 | | |
| 3 | | |
| 4 | | |
| 5 | | |
| 6 | | |
| 7 | | |
| 8 | | |
| 9 | | |
| 10 | | |
| 11 | | |
| 12 | | |
| 13 | | |
| 14 | | |
| 15 | | |
| 16 | | |
| 17 | | |
| 18 | | |
| 19 | | |
| 20 | | |

### Trajectory **B**

| DATA POINT | x [cm.] | y [cm.] |
|:---:|:---:|:---:|
| 1 | | |
| 2 | | |
| 3 | | |
| 4 | | |
| 5 | | |
| 6 | | |
| 7 | | |
| 8 | | |
| 9 | | |
| 10 | | |
| 11 | | |
| 12 | | |
| 13 | | |
| 14 | | |
| 15 | | |
| 16 | | |
| 17 | | |
| 18 | | |
| 19 | | |
| 20 | | |

| *Trajectory* **A** | *Trajectory* **B** |
|---|---|
| $x_1$ = _____ | $x_1$ = _____ |
| $y_1$ = _____ | $y_1$ = _____ |
| $x_2$ = _____ | $x_2$ = _____ |
| $y_2$ = _____ | $y_2$ = _____ |
| (at  x  =  14.0 cm.) | (at  x  =  14.0 cm.) |
| slope  =  _____ | slope  =  _____ |

# QUESTIONS:                 *Did you Learn Something ?*

> The horizontal intervals, $\Delta x$, are *directly* proportional to *time* - intervals, $\Delta t$, because :
>
> $$\Delta x = v_x \Delta t$$
>
> and
>
> $$v_x = (v_o)_x = constant$$

1.  If you think of your **x** -axis as your **time** axis , then what physical quantity does the slope of your tangent line represent ?

2.  Proceeding from left to right, and for equal horizontal intervals, $\Delta x$, are the vertical intervals, $\Delta y$, increasing, decreasing, or remaining the same ? Why do you think this is happening ?

3.  On the basis of your trajectory " A " : compare the slopes of tangential lines, for different
    data points:

    a)  Did the **velocity** of the ball change ?  Explain your answer.

    b)  Did the **horizontal component** of the **velocity** of the ball change ?  Explain.

    c)  Did the **vertical component** of the **velocity** of the ball change ?  Explain.

    d)  Did the  **speed** of the ball change ?  Explain your answer.

4.  Compare trajectory " A " with trajectory " B " at the x-displacement position of
    $$x  =  10.0 \text{ cm.}$$
    (Check one box)

    ☐     the accelerations are the same

    ☐     the accelerations on curve  A  is greater than the acceleration on curve  B

    ☐     the accelerations on curve  A  is greater than the acceleration on curve  B

    Explain your answer.

# " g "
# ACCELERATION OF GRAVITY

If you drop something, and your reflexes are slow, you might not be able to catch it. If a dime accidentally falls out of your hand while you're on the observation deck of the Empire-State building, it will only take about *10 seconds* to reach the ground !

The explanation is only partly due to the fast speed that the object has, because, after, it spends *most of its time* falling at smaller speeds ( starting from zero speed !). The object *accelerates*

What is just as interesting, is that objects gain or lose the <u>same amount of velocity</u>, during the <u>same time interval</u>, regardless of whether that time interval is near the beginning, the middle, or the end of the motion. The acceleration of freely falling objects, due to gravity, is *constant !*

What you're about to do, is explore a technique that will allow you to determine the specific constant value of the acceleration of gravity.

# The TOOL-BOX:

Air-Track          Photogate Timer          Vernier Caliper
Air Supply         Accessory Photogate      Meter Stick
Air Hose           Small Gold Glider

# STEP-BY-STEP :

1. Level the air-track by carefully and slowly adjusting the plastic wing nut under the support stand, while simultaneously observing the position of the glider. This is an extremely sensitive and accurate method of leveling. Although the glider may move slightly, due to air currents, it will not tend to continually drift in a specific direction, if the air-track is truly horizontal.

2. Using a Vernier Caliper, measure the height of a riser-block (a rigid flat block ), and record this in your data as **h** .

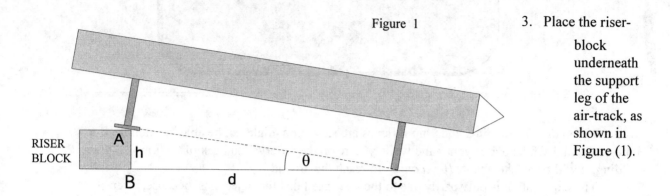

Figure  1

3. Place the riser-block underneath the support leg of the air-track, as shown in Figure (1).

RISER
BLOCK

A   h

B          d          C

θ

4. Using a meter stick, measure the distance between points  B and  C , as shown the Figure ; enter it as **d** in your data sheet.  Points  B  is where the front of the riser-block meets the table.  Point  C  is where the other support leg is in contact with the table.

5. Slide the Photogate Timer switch to  GATE  mode.

6. Measure  **L** , the length of the glider, in the same way that you did in the *Uniform Acceleration* experiment, i.e., by passing it, by hand, slowly through one of the Photogate Timers, and calculating the difference between where the glider makes the LED light go on and off, respectively.  Record this value in your data sheet.

7. Press the  RESET  button on the Timer.

8. Measure  **D**  , the distance between the two photogates, by holding the glider and slowly guiding it into the first Photogate.  When the RED light first comes on, note the corresponding scale position.  Now move the glider into the Accessory Photogate and note the scale position when the LED light goes on.  The difference between these scale positions is  **D** .

9. Press the  RESET button.

10. Place the glider near the top of the air-track, and record the scale position  $x_0$ , of the front of the glider.  Let the glider accelerate freely down the track.  One group partner has to *observe* , and *call out* the transit time  $t_1$  through the first Photogate, as well as press the RESET button *before* the glider passes through the Accessory Photogate, otherwise it will be *lost*.  A third group partner should be listening for, and writing down,  $t_1$  , as well as listening for, and writing down,  $t_2$  the transit time through the Accessory Photogate

11. Repeat steps  9  and  10 , three more times, so that you have a set of four trial runs.  Each partner in the group should perform a different operation, for each trial run.

# Show-the-World:

1.  In order to find the glider's acceleration, we first have to calculate the angle that the air-track is making with respect to the horizontal, as shown in Figure (1). Since the riser-block's height is **h**, and the horizontal distance **d** are the sides of the right triangle ABC, as shown in Figure (2), ,you can calculate the angle θ from:

Fig. 2

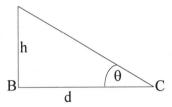

$$\theta = \tan^{-1}\left(\frac{h}{d}\right)$$

2.  Calculate the average transit times $t_1$ and $t_2$, from their respective set of four trial runs.

3.  Calculate the two velocities $v_1$ and $v_2$ that the glider had while passing through the two Photogates, by dividing the glider's length **L** by their respective transit times $t_1$ and $t_2$ :

$$v_1 = \frac{L}{t_1}$$

$$v_2 = \frac{L}{t_2}$$

4.  Calculate the glider's acceleration from:

$$v_2^2 - v_2^2 = 2\,a\,D$$

## *FINALLY !*

5.  To calculate **g**, the acceleration of gravity, which acts downward, you need to recognize that this is a vector that can be resolved into two component accelerations; one parallel to the air-track, and the other perpendicular to the air-track, as in Figure (3). But we're only interested in the *parallel* component, because that is the *actual acceleration of the glider !* You can calculate it by noticing that the angle θ also lies between **g** and the line perpendicular to the air-track, so that in this right triangle:

Solving this for **g**, will give you your experimentally determined value of it. Call it $g_{exp}$.

$$a = g \sin \theta$$

## *"WHAT IF I GOOFED?"*

6.   The way to tell is to compare your value with the reference value obtained by using more sensitive equipment.  For us, that would be:

$$g_{ref} = 981 \ cm/sec^2$$

A useful way of comparing the two, is to calculate the per-cent difference between them:

$$\% \ \text{difference} = \frac{g_{ref} - g_{exp}}{g_{ref}} \ 100 \%$$

# DATA:        *Measuring Mother Nature*

**d** = _____        **D** = _____        $x_0$ = _____

**h** = _____        **L** = _____

| | TIME TRIALS (sec) | | | | $t_{ave}$ | **V** (cm/s) |
|---|---|---|---|---|---|---|
| | 1st | 2nd | 3rd | 4th | | |
| $t_1$ | | | | | | |
| $t_2$ | | | | | | |

$\theta$ = _____                $a$ = _____

$g_{exp}$ = _____              $L$ = _____

$g_{ref}$ = __981  cm/sec$^2$__      % difference = _____ %
                                     (in $g$)

# QUESTIONS:          *Did you Learn Something ?*

1.  We assumed that the acceleration of the glider was constant.  Do you think that this was reasonable assumption to make ?  If so, why ? , If not why not ?  How would set up an experiment, using this equipment, to test how valid this assumption is ?

2.  Do you think the % difference would be the same if you were to do the same experiment, but on a 10-meter air-track ?  Explain you answer.

3.  What other factors do you think might have contributed to your % difference ?

4.  How would you design an experiment to obtain a very small  % difference ?

# FORCES in EQUILIBRIUM

**6**

How many forces are acting on you right now ?  One ?  Two ?,  More than two ?
Is there some way of knowing ?  For your rocket-ship to move from one point to another, in deep space, you don't have to burn any of your expensive fuel.  Moving at constant velocity, fortunately, is a "natural" thing to do, in the sense that it happens without making any special effort to cause it;  that is Newton's 1$^{st}$ Law.

Having *zero* velocity (being at rest), is just a special case of having constant velocity.  But if *any* force is acting, then there *must* be *at least one other* force acting to counter-balance it, so that the *net* force (the *resultant* force) is zero.  This *doesn't* mean that we now have a way of automatically telling *how many* forces are acting, but it does give us a clue as to whether the system is in  EQUILIBRIUM;  simply REMOVE one of the forces !  If the system was truly in equilibrium, it will now be unbalanced, and start to accelerate.  So, *one* force ( called the  "EQUILIBRANT" ) is responsible for, literally "holding the entire system together" .  But which one ?  The correct answer:  *Any* force can inherit the title of 'equilibrant'.  Any force can be the force which is equal in magnitude, and opposite in direction to the resultant (the vector sum) of all the other forces that may be there !

You're about to put these ideas to the experimental test, by finding the resultant of two forces, using the practical method of finding the equilibrant first.  While you're at it, you'll also explore how vector forces can be represented by two perpendicular components.

# The TOOL-BOX:

Force Table Assembly
(Force Table, Stand, Ring, Clamps with Pulleys)

Mass Holders
Various Masses
String

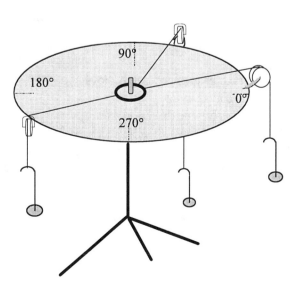

Figure 1

## *The Round-up !*                         *(for your convenience !)*

RESULTANT FORCE  ≡  the *vector* force which is a <u>*vector*</u> sum of *all* the forces acting.

EQUILIBRIUM  ≡  The condition a system is in when the *resultant force* on it is *zero*.

EQUILIBRANT  ≡  any force which is *equal in magnitude*, and *opposite in direction* to the
                           resultant of all the other forces acting.

RESOLUTION  ≡  the process of finding two perpendicular vectors to completely
                           represent the effect of any one vector.

UNCERTAINTY  ≡  the range of values, *above and below* your measured result, into which
                           you feel another measurement would *probably* fall.


## *INSPECTING THE MERCHANDISE !*

- Notice that the circular metal table is divided into 360 divisions, the smallest one equal to *one degree*.
- Each of the four pulleys can be moved to a different angle by simply loosening the wing-nut on the clamp, moving it, and retightening it.
- There is a center post, around which is a small ring, to which four strings are attached. Since each string can pass over a pulley, you can attach a mass holder at the string's other end, and place various masses on it. This givers you a convenient way of varying the amount of the tension force along the string !


# STEP-BY-STEP:

## Part I:      *The Resultant of 2 Forces*

1.  Measure the masses of three mass holders (they may not be the same!) using the balance.

2.  Allow three of the strings to pass over three pulleys.

3.  Move the clamps to the appropriate direction.

4.  Set up the two forces $F_A$ and $F_B$ , using the values given in the following chart, by
       placing the necessary masses onto two of the mass holders.  Keep in mind that the

> mass you need  =  mass of the holder  +  added mass

| Mass (grams) | Magnitude (Dynes) | Direction (degrees) |
|---|---|---|
| $m_A$ = 100 | $F_A$ = 98,000 | $\theta_A$ = 30° |
| $m_B$ = 150 | $F_B$ = 147,000 | $m_A$ = 130° |

$$F = W = m\,g \qquad\qquad g \text{ - } 980 \text{ cm/sec}^2$$

5. Start adding some masses onto the third mass holder, and hook it onto the third string.

6. Now try a combination of two things:
   - a) have someone in the group add masses to the third holder,
   - and  b) have someone in the group vary the angular direction of the third string, by moving the third clamp.

   When you have found those *unique* values of mass and direction that bring the system into equilibrium, record them for your equilibrant, in your Data sheet.

## Part II:   *Resolving Vectors into Perpendicular Components*

## A BRIGHT IDEA!

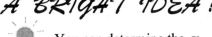

You can determine the **y** – component of $F_A$ : ( $F_{Ay}$ ) *experimentally* , if you know $F_A$ and ( $F_{Ax}$ ) .

1. First, calculate, *mathematically* , what the **x** – component of $F_A$ is, by using:

$$F_{ax} = F_A \cos \theta_A$$

2.    If you put $F_A$ and it's x– and y– components on the Force Table, they wouldn't balance ! But equilibrium would occur if you used the *equilibrant* of $F_A$ , (let's call it $E_A$ ) together with the components !  The equilibrant has the same magnitude as $F_A$ , but is *opposite in direction* , i.e., it lies in a direction 180° *from* the direction of $F_A$ .  Record the magnitude and direction of $E_A$ in your Data sheet

3.    Set up $E_A$ and $F_{ax}$ on your Force-Table.  Place $F_{ax}$ at 0° .  You are going to find $F_{ay}$ experimentally, so place the third string at 90° , to represent it.

4.    Vary the mass on the third string, similar to the way you did in step # 6 of Part I , in order to find the *unique* value of magnitude of the y- component, that brings the system into equilibrium.  Record this, in your data sheet, as your experimental value for the magnitude of $F_{ay}$ .

5.    Determine, experimentally, the *uncertainty* in *magnitude*, i.e., the *plus* and *minus* values, by using the same method that you used in steps # 7 - 8 of Part I .  Enter these two values in your Data sheet.

# Show-the-World:

1.    Since the tension forces in the string are due to the weights of the masses and mass holder, it's useful to be able to express everything as one or the other.  she "language of communication" between weight and mass is :

$$\boxed{W \;=\; m\,g}$$    where    $\overline{\quad g \;=\; 980 \ cm \,/\, sec^2 \quad}$

Make any necessary conversions.

## Part I

2.    Using Pythagoreans's Theorem, and Trigonometry, *calculate* the *magnitude and direction* of the Resultant Force., due to the *vector* addition of forces $F_A$ and $F_B$ .

 HINT:    ☞    | Draw a sketch of the forces, resolve them into their x- and y- components, enter them into the Data Chart , add all the x- components, add all the y- components, then calculate the magnitude and direction.

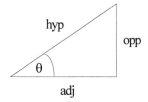

$$(hyp)^2 = (opp)^2 + (adj)^2$$

$$\sin \theta = \frac{opp}{hyp}$$

$$\cos \theta = \frac{adj}{hyp}$$

$$\tan \theta = \frac{opp}{adj}$$

3.   State the magnitude and *direction* of the resultant force you obtained *experimentally* , by having measured the *equilibrant* directly.

4.   Calculate the *per-cent difference* between the magnitude of you experimental value, as compared to your theoretical (mathematical)  value.

# DATA:       *Measuring Mother Nature*

mass of holders:       $m_1$  =  _____

$m_2$  =  _____

$m_3$  =  _____

## Part I

## EQUILIBRANT

m  (added)  =  _____

m  (TOTAL)  =  _____

*Magnitude*                                        *Direction*

E  =  _____ Dynes                    $\theta$  =  _____ °

# RESULTANT FORCE

### *Magnitude*                                   *Direction*

E  =  _____  Dynes                θ  =  _____ °

# UNCERTAINTY

### *Magnitude*                                   *Direction*

*Magnitude  PLUS  (+)*  =  _____        *Direction  PLUS  (+)*  =  _____ °

*Magnitude  MINUS  (−)*  =  _____       *Direction  MINUS  (−)*  =  _____ °

(Experimental vs. Mathematical)        % Error  =  _____ %

# Part II

(Mathematical)   $F_{Ax}$  =  _____              $\theta_{Ax}$  =  0.0 °

E$_A$  =  98,000 Dynes              $\theta_A$  =  _____ °

Experimental      m (added)  =  _____

m (TOTAL)  =  _____

Experimental      $F_{ay}$  =  _____              $\theta_{Ay}$  =  90 °

# UNCERTAINTY   in   $F_{ay}$

## *Magnitude*

*Magnitude  PLUS  (+)  =* _____

*Magnitude  MINUS  (−)  =* _____


# QUESTIONS:                *Did you Learn Something ?*

1.    What do you feel were the most important *sources* of error in this experiment.

You can run much faster if you don't have to carry 50 lbs. of camping equipment on your back. Exactly how does an object's inertia affect it's motion ? Many people have thought about this practical question, but it wasn't until 400 years ago that Galileo did an experiment to find out ! It was only 300 years ago when Isaac Newton followed up on Galileo's ideas, with his own experiments, and his conclusions changed the way we think about how nature works ! By eliminating friction, the one factor that makes it difficult to understand what's going on, you too will discover some of the natural laws of motion !

# The TOOL-BOX:

| | | | |
|---|---|---|---|
| Air-Track | Photogate Timer | Pulley | Sponge |
| Air Supply | Accessory Photogate | Assorted Masses | String |
| Air Hose | Large Red Glider | Mass Holder | |

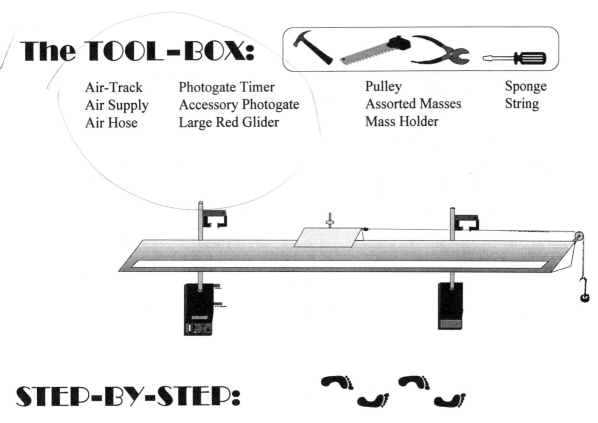

# STEP-BY-STEP:

1. Measure the mass of the glider, using the balance. Enter it as **m** in your data sheet.

2. Level the air-track carefully, in the usual way, using the glider as a detector.

3. Set up the equipment just as you see it in the Figure, by attaching a string between the glider and the mass holder, and allowing it to pass over the end pulley.

4. *Precisely* place the first Photogate at $s_o = 80$ cm. , and the second at $s_o = 125$ cm. ,

by:     a) Switching the Photogate to PULSE mode.
        b) Holding the glider at each location, then *slowly* move the Photogate until the
red LED light turns *on* . The difference of these two values is the separation distance $\Delta s$

$$\Delta s = s_f - s_o = 125 \text{ cm.} - 80 \text{ cm.} = 45 \text{ cm.}$$

# Part A:

(m$_{total}$ = constant)

To discover how acceleration depends on Net Force, you can change $F_{NET}$ by increasing $m_2$,

since:     $$F_{NET} = W_2 = m_2 \, g$$

*but* you must keep the *total mass* constant:     $$m_{total} = (m_1 + m_2) = \text{constant}$$

A1.   Start by placing 50 grams onto the glider (use 10 and 20 gram masses) , so that

$$m_1 = m_{glider} + 50 \text{ grams}$$

and     $$m_2 = m_{holder} + 0 \text{ grams}$$

A2.   Place the front of the glider *just before* the $s_o = 80$ cm. scale position. Place the sponge past
the $s_f = 125$ cm. scale position, so that the glider will hit the soft sponge, after passing $s_f$ ,
thereby preventing your experiment from **crashing apart** !

A3.   Set the Photogate Timer to PULSE mode, and press the RESET button.

A4.   Hold the glider as steady as you can (a ruler might help), and suddenly release it. After the glider
has passed completely through the Accessory Photogate, record the time displayed as $\Delta t$ .

A5.   Remove a 10 gram mass from the glider (leaving 40 grams *on* the glider), and place it on the
mass holder. By doing this, you have *increased* the *net-force* which accelerates the system,
while keeping the *total mass* of the system *constant*.

A5.   Remove a 10 gram mass from the glider (leaving 40 grams *on* the glider), and place it on the mass holder. By doing this, you have *increased* the *net-force* which accelerates the system, while keeping the *total mass* of the system *constant*.

A6.   Repeat the process, that you did in steps A2 through A4 , with this new arrangement of masses, and enter the time in your Data chart.

A7.   Perform *four more* experimental runs, each time transferring 10 grams from the glider to the holder. The final run should involve no masses on the glider, and 50 grams of mass on the mass holder.

# Part B:

## (Net Force  =  constant)

We're interested in discovering how a systems inertia affects it's acceleration. (Aren't you ?) But we have to make sure the Net Force which produces the acceleration is always the same value. It's *easy* to keep the Net Force constant, because the Net Force is just:

So just   $F_{NET}$  =  $W_2$  =  $m_2\,g$          keep  $m_2$  =  50 grams

In order to *vary* the *total mass* ,     $m_{total}$  =  $(m_1 + m_2)$

we can change $m_1$ by adding mass to the glider:

$$m_1  =  m_{glider}  +  \text{(mass on glider)}$$

B1.   Set the Photogate Timer to PULSE mode, and press the RESET button.

B2.   Place a 50 gram mass onto the mass holder.

B3.   For the first set of data, use *NO* mass on the glider, so that          $m_1$  =  $m_g$

B4.   Place the glider at $s_o$  =  80 cm., and release it. Record the time traveled between Photogates, and enter it in the $\Delta t$ column of Part B of your data sheet.

B5.   Now *add* a 10 gram mass to the top of the glider. The combined glider mass should now be

$$m_1  =  m_g  +  10 \text{ grams}$$     The mass of the holder is still 50 grams.

B6.    Repeat the experiment for this new mass combination, and record the travel-time and new masses in your data chart.

B7.    Perform the experiment *four more* times, each time adding 10 grams to the top of the glider, so that the masses you used for your six runs are as follows:

Run  1        $m_1$  =  $m_g$  +      0 grams
Run  2        $m_1$  =  $m_g$  +    10 grams
Run  3        $m_1$  =  $m_g$  +    20 grams
Run  4        $m_1$  =  $m_g$  +    30 grams
Run  5        $m_1$  =  $m_g$  +    40 grams
Run  6        $m_1$  =  $m_g$  +    50 grams

# Show-the-World:

1.  In Parts  A  and  B , we assumed that the acceleration of the glider was uniform (i.e., constant).  If this assumption is valid (to within a small experimental error), then you can calculate the system's acceleration form your experimentally determined values of  $\Delta s$ and   $\Delta t$  by using the equation of motion that doesn't have  velocity in it. specifically:

$$\Delta s  =  v_0 (\Delta t)  +  \tfrac{1}{2} a (\Delta t)^2$$

The system started  *from rest* , so,          $v_0  =  0$

The relationship simplifies to

$$\Delta s  =  \tfrac{1}{2} a (\Delta t)^2$$

where:          $\Delta s$  ⇝  glider travel distance
                    $\Delta t$  ⇝  glider travel time
                    $a$   ⇝  system's acceleration

Solve for the acceleration algebraically,

$$a  =  \frac{2 \Delta s}{(\Delta t)^2}$$

and use it to *calculate* the acceleration of each experimental run you performed, in Parts  A  and  B , using the corresponding values of  $\Delta s$  and  $\Delta t$ .  Enter your calculated acceleration values into your data chart.

2.   *Calculate* the force that the <u>mass holder</u> and it's <u>masses</u> applies to the system, by multiplying
this total mass ($m_2$) by the accepted value of the acceleration of gravity:

$g = 980 \text{ cm/sec}^2$ for each experimental run you performed, and enter it into your data chart.

$$F_{NET} = W_2 = m_2 g$$

3.   From your data in Part A, *plot a graph* of $F_{NET}$ vs. acceleration, putting Force on the
vertical axis, and acceleration on the horizontal axis.

4.   From your data in Part B, *calculate* the total system mass, by adding the total mass by adding
the combined mass (of the glider and it's masses), to the combined mass of the holder and its 50
gram mass).

$$m_{tot} = (m_g + m) + (m_g + m)$$

where:   $m_{tot}$ ⟿ total system mass

$m_g$ ⟿ total system mass

$m$ ⟿ total system mass

$m_h$ ⟿ total system mass

5.   From your data in Part B, *plot a graph* of acceleration vs. total system mass, putting
acceleration on the vertical axis and mass on the horizontal axis.

# DATA:        *Measuring Mother Nature*

( glider's mass only )   $m_g$ = _____ grams          $s_0$ = _____80___ cm.

( glider's mass only )   $m_g$ = _____ grams          $s_f$ = _____125___ cm.

$\Delta s = (s_f - s_0) =$ _____45___ cm.

## A ) <u>Acceleration</u>  vs.  <u>Net Force</u>

|  | $m_1$ (gms) | $m_1$ (gms) | $\Delta t$ (sec) | $F_{NET}$ (Dynes) | $a$ (cm/sec²) |
|---|---|---|---|---|---|
| $m_1 = m_g + 50$ |  |  | 1.02 |  |  |
| $m_1 = m_g + 40$ |  |  |  |  |  |
| $m_1 = m_g + 30$ |  |  |  |  |  |
| $m_1 = m_g + 20$ |  |  |  |  |  |
| $m_1 = m_g + 10$ |  |  |  |  |  |
| $m_1 = m_g + 0$ |  |  |  |  |  |

## B ) <u>Acceleration</u>  vs.  <u>Mass</u>

|  | $m_1$ (gms) | $m_1$ (gms) | $\Delta t$ (sec) | $F_{NET}$ (Dynes) | $a$ (cm/sec²) |
|---|---|---|---|---|---|
| $m_1 = m_g + 50$ |  |  |  |  |  |
| $m_1 = m_g + 40$ |  |  |  |  |  |
| $m_1 = m_g + 30$ |  |  |  |  |  |
| $m_1 = m_g + 20$ |  |  |  |  |  |
| $m_1 = m_g + 10$ |  |  |  |  |  |
| $m_1 = m_g + 0$ |  |  |  |  |  |

# QUESTIONS:          *Did you Learn Something ?*

1.  Does your graph for  Part  A  best fit a straight line or a curve ?

2.  Does your graph for Part  B  best fit a straight line or a curve ?

3.  Examine your graph of  acc.  vs.  mass  in the region where the masses you used are small.  Can you
    think of a practical reason why we chose to start Part  B  with at least a  100 gram mass added to
    the glider ?
    HINT:   Imagine the glider itself, to be only  ONE gram  of mass, and you were to carry out the
    experiment.  What do you think would happen ?

4.  Isaac Newton performed a third set of experiments related to what you've been doing, that enabled
    him to make a complete statement about the relationship of  net force, acceleration,  and mass.
    What was that third statement, and describe, in detail, how you would carry it out using the same
    equipment ?

Have you ever felt yourself going in circles ? Perhaps it was a Merry-Go-Round, or a Ferris-Wheel ? Or making a sharp turn in a car ? If you did, then you noticed the *apparent outward* pull on your body. *But it's just a figment of your imagination !* Your tendency was to go straight, as Galilee and Isaac Newton pointed out to us. It was actually the car or the Merry-Go-Round pushing on you in a direction perpendicular to your motion, and *in the direction of the turn*, that forced your body to deviate from that straight line. That's exactly what someone would see if they were looking at you from the ground. When the object pushes on you, you push back on the object (Newton's $3^{rd}$ Law), and this is what gives you the false sense of an outward force.

You also might have sensed that the faster you make the turn, the larger is the *center-directed* force (a literal translation of the word CENTRIPETAL) , but have you thought about just how large is it ?

You also probably realize that the "tighter" the turn, the greater the force. Olympic bobsled and Luge athletes are well aware of the consequences of misjudging the <u>speed</u> and the <u>radius</u> of the turn. What you're going to do is discover for yourself, by experiment, without needing to be an Olympic bobsleder (although one day you just might !) , just how centripetal force depends on the speed, and the radius of the turn. You'll be doing it by rotating a rubber plug with a string attached to it, by hand, in a circle, and clocking how long it takes to make a specific number of turns. You then have enough information to calculate the speed, centripetal acceleration, and centripetal force. You can compare this experimental value with the actual force that caused it: a hanging weight at the other end of the string.

# The TOOL-BOX:

| | | | |
|---|---|---|---|
| Large Rubber Plug | Tube Holder | Mass Holder | Timer |
| Small Rubber Plug | String | Assorted Masses | Meter Stick |

# STEP-BY-STEP:

## *The Setup*

1.  Measure the masses of the two rubber plugs.  Let's call the mass of the big plug **m$_B$** , and the small one **m$_S$** .

2.  Cut a piece of string to a length of about one meter, and tie one end of it to the large rubber plug.  Thread the string through the tube, and tie the other end to the small rubber plug.

3.  Lay the equipment on the table, and straighten it all out, as shown in Fig. (1).

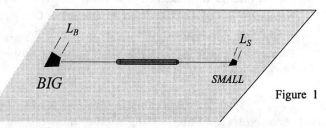

Figure  1

4.  Measure the length of each rubber plug.  Let's call them **L$_B$ and L$_S$** .  You'll do the first part of the experiment using a radius of  60 cm.  To get a reasonably accurate length of string, subtract *half* of L , from 60 cm.  This is the length of string that should lie between the top of the tube and the bottom of the small plug (its the *small*  plug that you'll be rotating around in a circle).

$$\text{distance} \;=\; 60\,\text{cm.} \;-\; \left(\frac{\mathbf{L_S}}{2}\right)$$

5.  Using a pen, make a noticeable mark on the string at a point about  2 centimeters *below* the bottom of the tube()on the big plug side of the string), as shown in the Figure.

## ✌ *The Game Plan !*

     One team member will be concentrating on rotating the small plug by holding the tube and flexing their wrist, always sure that the radius never changes.  Another team member will be watching the plug go around, counting the number of complete revolutions made, and controlling the timer.  The weight of the large rubber plug produces the tension in the string that supplies the needed centripetal force, as shown in Figure (2).

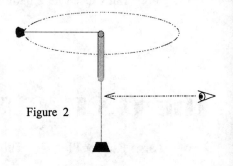

Figure 2

## *Practice Makes Perfect !*

     Making the plug go around in a horizontal circle at a constant speed, looks easier than it is to do it !  Everyone in the group should practice doing it *several times*, until everyone feels comfortable with it.

# *The TRICKS of the TRADE !*

✔ Grasp the rubber tube on the rod, making sure your hand is not touching the string.

✔ If you're the person doing the spinning, then you should extend your hand *all the way out*, and at an *upward angle* , to avoid hitting yourself (or anybody else) in the head.

✔ Hold the *bigger mass* with your other hand, and start the motion by rotating your hand only, *not you arm.*

✔ Always keep the rod in a *vertical* position.

✔ Release your hand from the larger plug.

✔ Let another team member *briefly* , and *gently* , touch the bigger hanging mass to stop it from swinging.

✔ Spin it just fast enough so that the small plug rotates in a *horizontal* plane.

✔ Concentrate on the mark you made below the tube.  Try to always maintain the same distance between that mark and the bottom of the tube.

✔ OK, now imagine that you've done a successful run.  How are you going to stop it ? Ducking your head out of the way, is *dangerous !*  It's much easier and safer to just pull the larger mass *down* with your other hand !  By doing this, you'll be decreasing the radius, until the small plug reaches the tube, causing it to stop.  But look what happens to the speed of the small plug when you're drawing it in;  it speeds up !  We'll talk about this later.

✔ Most importantly, *DON'T GET DISCOURAGED !*  You can do it !  It just takes practice.  You never know when success is just around the corner !

# *LIGHTS ! CAMERA ! ACTION !*

1. When the partner  A  (who's doing the rotating), feels comfortable, he/she should call out "READY" to the other partner.

2.  Partner B should look at a fixed reference point in the background, that the rotating plug passes over, to make it easy to count complete revolutions, and begin counting for **40 revolutions**.  Press the start button on the Timer simultaneously with the *beginning* of the first turn.

3.  When exactly  40  complete revolutions have gone by, partner  B  stops the Timer, and calls out "STOP" , at which time partner  A  performs the stopping routine.  Record the time in your data chart.

4.  Repeat the process twice more, for a total of three trial runs.

5.  Repeat the SETUP procedure once more, but this time arrange it so that the radius of the circle will be **30 cm.** , instead of  60 cm.

6.  Perform the experiment for three trial runs with the new radius.

# Show-the-World:

1. Calculate the average of your measured times, for a set of three trial runs, by adding the times and dividing by three:

$$t_{ave} = \frac{t_1 + t_2 + t_3}{3}$$

2.  Calculate the **period (T)** of a  revolution (the time it takes to go once around), by dividing your average total times by   N = 40  , the number of revolutions:

$$T = \frac{t_{ave}}{N}$$

3.   Now, you have enough information to calculate the speed of the small rubber plug.  The distance it travels during the course of one revolution is the  **circumference (C)** of the circle:  $C = 2\pi r$ , and the time it took to go around once is the  **period (T)** , so:

$$v = \frac{C}{T} = \frac{2\pi r}{T}$$

4.   The expression for the **centripetal acceleration** is :

$$a = \frac{v^2}{r}$$

Calculate it for all your data.

5.  The **centripetal force**, which is the resultant force on the small cork, is the product of the mass of the *small* cork, and the centripetal acceleration:

$$F = m_S \, a$$

Calculate it for all your data.

6.   But in your experiment there is only one force acting on the small plug along the radial direction, and that is the tension in the string.  But the tension is being generated by the weight of the *big* plug:

$$W_B = m_B \, g$$

Calculate it for all your data.

7.   You now have *two* different ways of evaluating the centripetal acceleration.  One method is based on the data from your experiment, culminating in your calculation for step #5 ; let's call it the <u>experimental value</u>:   $F_{exp}$ .
The other method is the direct calculation of the weight of the big mass, in step #6 .
Let's call it the <u>theoretical value</u>:   $W_B$ .Make a comparison of the two by calculating the  % difference, with respect to the theoretical value:

$$\% \text{ difference} = \frac{W_B - F_{exp}}{W_B} \ 100 \, \%$$

# DATA:          *Measuring Mother Nature*

$m_S$ = _____          $m_B$ = _____

$L_S$ = _____          $L_B$ = _____

$N$ =   40 rev.          $W_B = m_B \, g$ =

_____

| $r$ (cm) | $t_1$ (sec) | $t_2$ (sec) | $t_3$ (sec) | $t_{avg}$ (sec) | $T$ (sec) | $v$ (cm/sec) | $a$ (cm/sec$^2$) | $F_S$ (Dynes) |
|---|---|---|---|---|---|---|---|---|
| 60 | | | | | | | | |
| 30 | | | | | | | | |

% difference = _____ %

# QUESTIONS:          *Did you Learn Something ?*

1.  Why was it important that the hand of the person doing the spinning didn't touch the string ?

2.  Why do you think the small plug sped up when the string was pulled to stop it at the end of the run ?
    HINT: Think about the physical quantities that the centripetal force depends upon.

3.  Why would it <u>not</u> have been a good idea to spin it for one revolution, instead of 40 revolutions ?

    Do you think that your % difference would have been larger, or smaller, if you had rotated it for 100 revolutions, instead of 40 ? Explain your reasoning.

4.  When you started to spin it faster and faster, the radius *increased*.  Why do you think this
    happened ?

5.  You were asked to keep it rotating in a horizontal plane.  If the
    string was making angle  θ , below the horizontal, i.e.,
    sweeping over the surface of an imaginary cone (*a conical-
    pendulum*),   why would the radius you measured, and used in
    your calculations, lead to a larger error ?
    HINT:   Look at Figure (3) , and draw a Free-Body Diagram of
            the forces acting on the small plug.

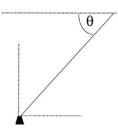

Figure  3

Pinball machine handles, bathroom scales, shock absorbers, and running shoes all have one thing in common: they are ELASTIC ; they can be stretched or compressed. A spiral spring is also elastic, and if it's not stretched or compressed *too far*, it will return to it's original shape and position (i.e., it's *equilibrium position*) when you take away the force that distorted it. But in 1678 , the English physicist Robert Hooke discovered experimentally another interesting phenomena: it takes *twice* as much *force* to stretch a spring to *twice* the stretch distance, *three* times the force for *three* times the stretch, and so on. This linear relationship isn't obvious at all, but you can repeat his simple experiment, and come to your own conclusion. Some springs are "stiffer" than others; a car's shock absorber spring will compress only millimeters, even though it takes *thousands of pounds* to do it ! You can determine the stiffness of your spring from *your own* experimental results !

# The TOOL-BOX:

Hooke's Law Apparatus (Stand, Spring, Mirror Scale, Mass Holder)

Various Masses

## WHAT'S WHAT ?

Some definitions will come in handy at this point:

**_Equilibrium Position_**  $\equiv$  The position that the spring is in, when it is balanced (not accelerating).

**_Elongation_**  $\Delta x$  $\equiv$  The distance that the spring stretches or compresses from it's previous equilibrium position.

**_Spring Constant_** (*k*)  $\equiv$  The stiffness of the spring. The ratio of applied force to elongation  $\Delta x$ .

**_Restoring Force_**  $\equiv$  The force that the spring exerts back on the attached mass. The restoring force is *always opposite* to the direction of the  *displacement* of the mass.

## *BEWARE OF THE PARALLAX !!!*

You'll be able to read the equilibrium positions of the spring by looking at the indicator needle attached to it, and compare it to the scale on the main support stand, whose divisions in centimeters and millimeters. A *source of serious* error can occur if you happen to be looking in a direction which is not horizontal, as shown in Figure (1). This parallax error can be avoided with the help of the mirrored surface of the scale. All you need to do is make sure that you look in the direction that allows you to see the indicator needle lined up with it's reflection, as in Figure (2).

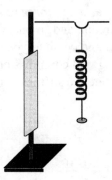

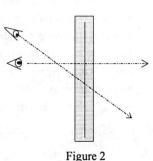

Figure 2

Figure 1

# STEP-BY-STEP:

## Part A:  *HOW STIFF IS IT ?*

A1.    Slide the mirrored scale up or down until the pointer is aligned with zero mark on the          scale. The current equilibrium position is now:          $x_0 = 0$ .

A2.    Place a  150 gram mass on the mass holder, and record the new equilibrium position as  **x** .

## Part B:  *HOOKE'S LAW*

B1 '    Remove the masses, and make sure the pointer is aligned with the zero scale position.

B2.    Place a  20  gram mass onto the mass holder, and accurately record the pointer's position after the spring has reached equilibrium.

B3.    Add another  20  gram mass onto the holder, giving a total of  40 grams, and record the pointer reading at the new equilibrium position.

B4.    Repeat the process, adding a  20  gram mass each time until you've generated <u>ten</u> sets of data:
      **m**  =  0, 20, 40, 60, 80, 100, 120, 120, 140, 160, and 180 grams .

# Part C:  [ *A STRING of SPRINGS* ]

What happens when two springs are connected together in series, and stretched by a hanging weight ?  Is the total elongation simply equal to the sum of the individual elongations, when acting alone, or is the situation more complicated ?

What's your opinion, and why ?  _____

_____

_____

_____

Let's perform an experiment to test your hypothesis !

C1.   *Very gently*, loop the top end of your spring onto the end of an *identical* second spring, and mount the combination back in place, as shown in Figure (3) .

C2.   Slide the mirrored scale down on the post until the pointer coincides with zero position.

C3.   Add a  20 gram mass to the holder, and record the needle's new equilibrium position.

C4.   Repeat step  C3  until you've obtained data for masses of:   0, 20,  40,  60,  80,  and  100 grams.

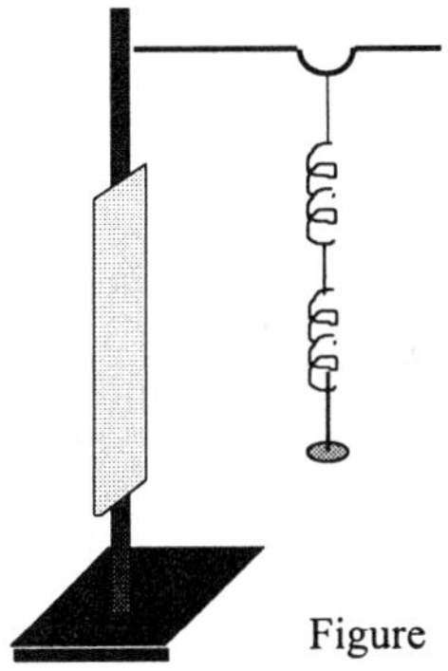

Figure 3

# Show-the-World:

# Part  A:  [ *HOW  STIFF  IS  IT ?* ]

1.  The weight of the hanging mass is

$$\mathbf{W} \ = \ \mathbf{m\,g}$$

In your experiment    m   =   150 grams .

This was the force which stretched the spring, and, since the spring was in equilibrium, it is also equal to the restoring force on the spring.  If Hooke's Law were valid for your spring, then:

$$\mathbf{F} = k\,\Delta\mathbf{x}$$

where

$$\Delta\mathbf{x} = (\mathbf{x} - 0) = \mathbf{x}$$

so:

$$m\,g = k\,\mathbf{x}$$

Since you already know   m , g ,  and  **x** , you can solve for the spring constant  k :

$$k = \frac{m\,\mathbf{g}}{\mathbf{x}}$$

Calculate the spring constant  k .

# Part B:  *HOOKE'S LAW*

2.  *Plot a graph* of   **F**   vs.   **x** , with  **F**  on the vertical axis, and  **x**  on the horizontal axis.

3.  Draw a  *straight line*  that  *best fits*  all of your data points from  B4 .

4.  choose two points  *on*  your straight line, and measure their coordinate values:
$$(\,\mathbf{F}_1\ ,\ \mathbf{x}_1\,)\quad\text{and}\quad(\,\mathbf{F}_2\ ,\ \mathbf{x}_2\,)$$

5.  Calculate the slope of your line:

$$\text{slope} = \frac{rise}{run} = \frac{\mathbf{F}_2 - \mathbf{F}_1}{\mathbf{x}_2 - \mathbf{x}_1}$$

# Part C:  *A STRING of SPRINGS*

6.  Repeat the type of calculation you did for Part A , but use, instead, the data you obtained for the two springs in *series* .  Label the spring constant , here:   $k_C$ , as though we were representing the two by one giant spring whose stiffness is  $k_C$ .

# DATA:          *Measuring Mother Nature*

## Part A:

$m = \underline{\phantom{xx} 150 \text{ gms} \phantom{xx}}$

$g = \underline{\phantom{xx} 981 \text{ cm/sec}^2 \phantom{xx}}$

$W = \underline{\phantom{xxxxxx}}$

$k = \underline{\phantom{xxxxxx}}$

$x_o = \underline{\phantom{xxxxxx}}$

$x = \underline{\phantom{xxxxxx}}$

$x_1 = \underline{\phantom{xxxxxx}}$

$F_1 = \underline{\phantom{xxxxxx}}$

$x_2 = \underline{\phantom{xxxxxx}}$

$F_2 = \underline{\phantom{xxxxxx}}$

slope $= \underline{\phantom{xxxxxx}}$

## Part B:

| m (gms) | x (cm.) |
|---------|---------|
| 0 | |
| 20 | |
| 40 | |
| 60 | |
| 80 | |
| 100 | |
| 120 | |
| 140 | |
| 160 | |
| 180 | |

## Part C:

| m (gms) | 0 | 20 | 40 | 60 | 80 | 100 |
|---|---|---|---|---|---|---|
| x (cm) | | | | | | |

# QUESTIONS:           *Did you Learn Something ?*

1.   Do you think a straight line was the best fit to your data points in your graph for Part B ?

2.   What are the units for the slope of your graph ?

3.   What physical quantity do you think the slope of the graph represents ?  What do you base your answer upon ?

4.   Compare the slope of your graph to the value of your spring constant from Part A , by calculating the % difference between them, with respect to your spring constant from Part A:

$$\% \text{ difference} = \frac{k_A - \text{slope}}{k_A} \, 100 \%$$

5.  What problems do you consider to be mainly responsible for generating this % difference ?
    Explain your reasoning, in detail.

6.  Compare the value of $k_C$ which you obtained for the two series-connected springs in
    Part C, with the *sum* of the spring constants of each single spring you determined from Part A .

    $k_C$  =  _____          $k_A$  +  $k_A$  =  _____

    Is your hypothesis correct ?

    If your answer is  YES , then how would you extend the experiment to give it a firmer base of
    support ?  If your answer is  NO , then try to come up with your own mathematical combination
    to account for your result, and try to give it some physical justification .

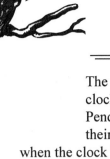

The pendulum is a fascinating device, whether it's the one inside your grandfather's clock, or inside the imagination of Edgar Allan Poe's story "The Pit and the Pendulum". An athletic coach will watch his runner's swinging arms, to improve their pendulum motion. It's used inside Grandfather-Clocks to compensate for when the clock "speeds up" , or "slows down".

But of all the varied and sometimes complex forms that you could design a pendulum, or a system of pendulums, to be, the simplest pendulum could be nothing more than a piece of string and a small object hanging from it ! The physics that you need to understand it's motion, you've already learned, yet many people are surprised when their hypotheses about its motion turn out to be wrong, so let's have some fun with the "Simple Pendulum" !

# The TOOL-BOX:

| | | |
|---|---|---|
| String | Assorted Masses | Photogate Timer |
| Rod Stand | Meter Stick | Masking Tape |

## " What's Your Name ? "

There is a convenient "language" associated with motion like that of a pendulum:

PERIODIC MOTION ≡ Any phenomena that repeats itself at regular intervals.
CYCLE         ≡ The basic repeating pattern of the periodic motion.
PERIOD      ≡ The amount of time it takes to complete one cycle.
FREQUENCY  ≡ The number of cycles completed in *one second* (1 Hertz ≡ 1 cyc/sec)
AMPLITUDE   ≡ The maximum swing of the pendulum, as measured by the angle
                      between the pendulum and a vertical line through its pivot point.

### *WHAT DO YOU REALLY BELIEVE ?*      *(Circle your answer)*

1. If the length of a pendulum were made *longer*, the *period* would be:
    a) the same             b) larger             c) smaller

2. If the length of a pendulum were made *shorter*, the *frequency* would be:
    a) the same             b) larger             c) smaller

3. If the length of a pendulum were made *longer*, the *maximum speed* would be:
    a) the same             b) larger             c) smaller

4. If the mass hanging from the end (the "bob"), were made *heavier*, the *period* would be:
    a) the same             b) larger             c) smaller

5. If you start the motion off with a *large amplitude* (large angle), the *period* would be:
   a) the same                 b) larger                 c) smaller

6. If you swung the pendulum on the surface of the Moon, the *period*, as compared to swinging it on Earth, would be:
   a) the same                 b) larger                 c) smaller


Let's perform some scientific experiments to check-out your hypotheses !


# STEP-BY-STEP:

## Part A:      *Period*

A1.  Cut a 60 cm. length of string and tape one end of it to a 10 gram mass.

A2.  Suspend the other end of the string from a point on the extension arm of the stand in a way that allows it to swing freely, as shown in Figure (1) . If possible, suspend the pendulum from two points, instead of one.  This prevents "wobbling", by forcing it to swing in a flat plane.

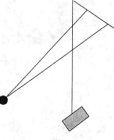

Figure 1

A3.  Invert the "U" shape of the Photogate and slide it far enough down its post so that it lies directly underneath the hanging mass, as shown in Figure (2) .

Figure 2

A4.  Set the Photogate Timer to GATE mode, and slide the Photogate up or down, in order to make sure the interrupts the beam.

A5.  Slide the Photogate Timer switch to PENDULUM mode.

A6.  Hold the suspended mass to the side, at an angle of about 20° , and let it go.

A7.  The Timer will temporarily stop counting at the end of the first cycle, but will resume counting when the mass next passes through the Photogate to begin the next cycle. *Before* it has a chance to pass through on the 1st quarter of the 2nd cycle, press the RESET button, and record the period as **T** , in your data chart.

A8.  Repeat the measurement of the period, step A7 , over 3 (three) later cycles. *Be careful not to interrupt the swinging !*

## Part B: *ANOTHER  FINE  **Mass** YOU'VE GOTTEN US INTO !*

B1.  Remove the  10 gram mass, and carefully replace it with a  20 gram mass.

B2.  Press the RESET button on the Photogate Timer.

B3.  Repeat the process you did in Part A , so that you're able to obtain four (4) consecutive measurements of the Period. *It's important that you don't interrupt the swinging !*

## Part C: *YOU'LL GO TO ANY **Length**, WON'T YOU ?*

C1.  Cut  4 more pieces of string, of lengths;  20 cm.,  40 cm.,  80 cm.,  and 100 cm..

C2.  Mount the other end onto the extension rod of the support stand, as you did previously.

C3.  Repeat the experiment by timing the periods of four (4) successive cycles for each length.

## Part D: *TURN UP THE **Amplitude**, PLEASE !*

This the experiment, whose results you're going to use to evaluate your hypothesis in Question # 5 .

D1.  Tape the  20 gram mass to the end of the  60 cm. string, and mount it as in previous experiments.

D2.  Raise the Bob (the mass) up and out until the string is horizontal, i.e., until it is making an angle of  90° with the vertical.  Make sure that you are holding the string taut.  Then release it.

D3.  Repeat the measurement process, in the same way you've done before, recording a set of four (4) trial times, without interrupting the motion of the pendulum.

## Part E: *BANG ! ZOOM ! STRAIGHT TO THE **Moon** !*

Oh! , by the way, about the Moon experiment:  you'll have to postpone it;  the school's budget can't afford to pay N.A.S.A. for a space flight to the Moon !  But, stay optimistic!, there are other ways of testing the validity of your hypothesis in Question #6.   For example, if you go to the lobby of the Hayden Planetarium, in New York City, you'll see 10 scales, representing the nine planets and the Moon.  You can stand on them and discover what you weigh on each.  You'll find, perhaps to you astonishment, that your weight on the Moon is only about  1/6 of your Earth weight ( A quick way to lose weight  ! ).

A similar situation (but to a lesser extent) exists right here on Earth.  Your weight at high altitude is actually different than at a low altitude ! (Prove it to yourself !)  You can use this weight variation to help you in you comparison of the Moon with Earth.  The next time you are in a mountainous area, at a high elevation, make a simple pendulum, and, with a stop-watch, time 20 cycles.  Divide this time by 20, to give you the average period (T).  Save the string and mass, and do the exact same experiment, the next time you're below sea-level.

# Show-the-World:

1.  Calculate the  AVERAGE PERIOD  from each set of four trial runs for the Period:

$$T = \frac{T_1 + T_2 + T_3 + T_4}{4}$$

2.  From your data in Part C, *plot two graphs*.  On one graph, put the Period **T** on the vertical scale, and put the length **L** on the horizontal scale.  On your 2$^{nd}$ graph, scale your vertical axis for the *square of the Period*, **T**$^2$, and still use the length **L** for your horizontal axis.  The reason we're scaling the horizontal axis as length, is that it was the *independent* variable in your experiment.

# DATA:          *Measuring Mother Nature*

| | $T_1$ | $T_2$ | $T_3$ | $T_4$ | $T_{avg}$ | $T^2$ |
|---|---|---|---|---|---|---|
| **Part A PERIOD** — m = 10 gm | | | | | | |
| **Part B MASS** m = 20 gms. θ = 20° — L = 60 cm | | | | | | |
| L = 20 cm | | | | | | |
| **Part C LENGTH** — L = 40 cm | | | | | | |
| L = 80 cm | | | | | | |
| L = 100 cm | | | | | | |
| **Part D AMPLITUDE** — θ = 90° | | | | | | |

# QUESTIONS:                *Did you Learn Something ?*

1.  If you allowed your pendulum to swing many times, you might have noticed it's Amplitude gradually decreasing with each cycle until, eventually, it would come to rest.  What do you think caused this to occur ?

2.  <u>MASS</u>:  In comparing your data from Part  B , with that of  Part A , what do you conclude about the effect of doubling the mass of the bob, on the Period,  **T** , of the motion ?

3.  <u>LENGTH</u>:
    a)  Is your graph of  **T**  vs. **length**  a straight line or a curve ?
    b)  Is your graph of  $T^2$  vs. **length**  a straight line or a curve ?
    c)  What do you conclude about how the length of the pendulum affects it's Period ?

4.  <u>AMPLITUDE</u>:  Compare your data from Part C, with that of Part B, where you kept the mass the same.
    a)  What conclusion do you come to, concerning the effect of varying the Amplitude, on the Period ?
    b)  Can you explain, based on your knowledge of Physics, why this might occur ?

5.  Discuss what you feel were the main sources of error in the experiments you performed.

## CONSERVATION of MOMENTUM

**11**

When objects bounce off each other, they may not change the direction they were moving in.  We certainly know what happens to a set of pins after being struck by a bowling ball, but if you look carefully the next time you go bowling, you might see the bowling ball deflected into a different direction by  the pins !

Mother Nature seems to assign a special vector to each moving object, that we humans call MOMENTUM.  An object's momentum is completely determined by the product of its mass and its velocity (the direction of its momentum is the same as the direction of its velocity).  What you're about to discover experimentally, is that when you add the momentum of all of the objects, *before* the collision, you get *the same answer* (to within a small experimental error) as what you get when you add the momentum of all of the objects *after* the collision.  This extremely important game that Nature plays, we call *Conservation of Momentum,* and it, along with other Conservation laws helps to make physics *the most fundamental* of all the sciences, and ushered in the modern age of technology.

# The TOOL-BOX:

| | | |
|---|---|---|
| Air-Track | Photogate-Timer | 2 Gliders |
| Air Supply | Accessory Photogate | Metal Bumpers |
| Air Hose | Electric Glider Launcher | |

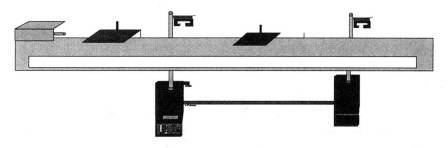

Figure 1

# STEP-BY-STEP:

1. Carefully level the air-track..

2. Set up the equipment as shown in Figure (1) .

3.  Measure the lengths, $L_1$ and $L_2$ , of the two gliders by the precise method we described using a *single* Photogate, in Experiment # 3 , "Uniform Acceleration", Step # 3, called "On Your Mark" :

 *On your Mark ....!*

> While holding the glider, *v-e-r-y s-l-o-w-l-y* guide it through the Photogate, and record the scale position, $L_0$ , at which the LED light on the Photogate Timer first goes on.  Continue guiding the glider until the LED light goes off, and record the location of the front of the glider as $L_1$ .  The difference of these two values is the effective length of the glider.                $L = L_{ON} - L_{OFF}$

4.  Weigh the mass of each glider: $m_1$ and $m_2$ , and enter these in your data sheet.

5.  Switch the Photogate Timer to GATE mode, and press the RESET button.

6.  Place glider 2 *in-between* the two Photogates.  Since this glider is at *rest*, it has zero velocity, and therefore has <u>zero momentum</u>.  Place glider 1 against the Electric Glider Launcher;  it has a small magnet that will hold the glider to it until its ready to be launched.

### COMING ATTRACTIONS !

Let's preview what will happen next, so that each group member will know exactly what he or she should do.  When the button on the Electric Glider Launcher is pressed, a plunger rod pushes out against glider 1 , thereby giving it an original velocity $v_{1o}$ , and a corresponding momentum $p_{1o} = m_1 v_{1o}$ .  It will then pass through Photogate 1 , registering a transit time $t_{1o}$ .  Next, it will crash into glider 2 , causing glider 2 to pass through Photogate 2 with a transit time of $t_{2f}$ .  Glider 1 , meanwhile,  will rebound back through Photogate 1 in a transit time of $t_{1f}$ .

### THE GAME PLAN !

The events occur so quickly, that it helps to be organized about *who* does *what*, and *when*.  One team member should control the Electric Glider Launcher.  Another person should observe and record the transit times, and then press the RESET button *before* either glider can pass into either Photogate.  Another partner should watch the gliders, to make sure they don't trip the Photogates more times than necessary, and also to stop them after the observations are made.

Let's summarize the transit times that have to be recorded:

$t_{1o}$ ≡ the time **glider 1** takes to pass through **Photogate 1** **before** the collision
$t_{1f}$ ≡ the time **glider 1** takes to pass through **Photogate 1** **before** the collision
$t_{2f}$ ≡ the time **glider 2** takes to pass through **Photogate 2** **before** the collision

## *PRACTICE MAKES PERFECT !*

The group should practice the sequence of events *several times* , until everybody feels confident enough to make the final run.

⇨  *GO FOR THE GOLD !!*

# Show-the-World:

**\*\*\* *REMINDER* \*\*\***

Velocity and momentum have directions, and are therefore **VECTORS** . You have to assign a sign to them. Let's call positive (+) , the direction of glider 1 before the collision, so that anything in the opposite direction is negative (–) .

1.  Calculate the velocities of the gliders, both before and after the collision: (use either a (+) or a (–) sign)

$$v_{1o} = \frac{L_1}{t_{1o}} \qquad\qquad v_1 = \frac{L_1}{t_1}$$

$$v_{2o} = 0 \qquad\qquad\qquad v_2 = \frac{L_2}{t_2}$$

2.  Calculate the momentums of the gliders, both before and after the collision:

$$p_{1o} = m_1 v_{1o} \qquad\qquad p_{1f} = m_1 v_{1f}$$

$$p_{2o} = m_2 v_{2o} \qquad\qquad p_{2f} = m_2 v_{2f}$$

3.  Calculate the total system's momentum before the collision, as well as the total system's momentum after the collision:

$$P_{tot} \ \text{before} = p_{1o} + p_{2o}$$

$$P_{tot} \ \text{after} = p_{1f} + p_{2f}$$

4. Calculate the per-cent difference between the total momentums, with respect to the total momentum before the collision:

$$\% \text{ difference} = \frac{P_{tot} \text{ after} - P_{tot} \text{ before}}{P_{tot} \text{ before}} \, 100 \%$$

# DATA:     *Measuring Mother Nature*

$m_1 = \underline{\hspace{3cm}}$          $L_1 = \underline{\hspace{3cm}}$

$m_2 = \underline{\hspace{3cm}}$          $L_2 = \underline{\hspace{3cm}}$

$v_{2o} = \underline{\hspace{1cm} 0 \hspace{1cm}}$          $t_{1o} = \underline{\hspace{3cm}}$

$v_2 = \underline{\hspace{3cm}}$          $t_1 = \underline{\hspace{3cm}}$

                                                    $t_2 = \underline{\hspace{3cm}}$

$p_{10} = \underline{\hspace{3cm}}$          $p_{1f} = \underline{\hspace{3cm}}$

$p_{2o} = \underline{\hspace{3cm}}$          $p_{2f} = \underline{\hspace{3cm}}$

$P_{tot} \text{ before} = \underline{\hspace{3cm}}$          $P_{tot} \text{ after} = \underline{\hspace{3cm}}$

$\% \text{ difference} = \underline{\hspace{3cm}}$

# QUESTIONS:     *Did you Learn Something ?*

1.  Was momentum conserved in your collisions?   If not, try to explain those factors that you feel contributed to the discrepancies.

2.  If a glider collides with the end of the air-track and rebounds, it will have nearly the same momentum it had before it collided, but in the opposite direction.  Is momentum conserved in such a collision ?  Explain.

3.  Suppose the air-track was tilted during the experiment.  Would momentum be conserved in the collision ?  Why, or why not ?

' Is it soup yet ? ' Why does it take more time to heat a quart of soup than it does a pint of the same soup ? After all, they start out at the same temperature, and they're heated to the same final temperature, so the temperature *change* they experience is the same ! If you answered "the quart has more volume therefore it has more mass" (mass = Vol. x Density), you'd be correct but you still haven't answered the question. A related question is: Why is the *inside* of a freshly baked chicken pot pie *hotter than the crust ?*

The answers to these, and similar questions, are based on the fundamental difference between *temperature*, and *heat*. While *temperature* is *related* to the average kinetic energy of the molecules, *heat* is actually a form of random *energy, transferred* between two points *only when a temperature difference exists !*

It actually takes about *three times as much* thermal energy to warm a copper penny by 1°C than it does the same mass of dirt ! The amount of heat gained or lost during the same temperature change varies from substance to substance because of the different molecular structures. The ratio of Heat (Q) to Temperature *change* (ΔT), is called the *Specific Heat Capacity*. It's nickname is *Specific Heat*, and you're about to discover the specific heat of a metal !

# The TOOL-BOX :

Aluminum Calorimeter Set:

```
⎛ Outer Can, Inner Can,              ⎞
⎜ Insulating Ring, Stirrer, Cover,   ⎟
⎝ Rubber Stopper w / Hole            ⎠
```

2　0°–100° Celsius thermometers　　2–3 Metal Rods
Bunsen Burner with Hose　　Mass Balance
Stand, Clamp, Heating Pad　　String
Copper Boiler　　Scissors

　　+　　　　=　　

*cool water*　　　　*hot metal*　　　　*equilibrium*

## What's the Recipe ?

✳ You can find the *specific-heat* of a metal by first *"cooking"* it in boiling water, then *"bathing"* it in cooler water, until it reaches *thermal equilibrium.* By measuring your masses and temperatures, the Heat exchanged can be calculated from:

$$Q = m\,c\,\Delta T$$

where　c ⇨ specific heat

　　　m ⇨ mass

　　　ΔT ⇨ change in temperature

By applying Conservation of Energy in the form of :

| HEAT LOST = HEAT GAINED |

you can determine how much heat is lost by the metal.

You can then find $c_{metal}$ by applying $Q = m\,c\,\Delta T$ to the metal.

# STEP-BY-STEP:

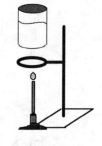

1. Fill the Copper boiler to about 75 % full of warm tap water, and place it on the heating pad near the top of the support stand.

2. Measure, and record in your data sheet, the mass of
    a) the empty inner calorimeter can,
    b) the stirrer
    c) the metal rods.

3. Tie the metal rods together with a piece of string, and drop them into the water. You can tie the other end of the string to a higher point on the stand, or just let it hang over the side of the boiler. Insert the first thermometer into the water.

Figure 1

4. Connect the Bunsen Burner and hose to the gas outlet, and center it under the stand (Figure 1). Turn the handle of the gas outlet, and light the gas with a sparker or a long match.
    **BE CAREFUL ! USE COMMON SENSE !** Make sure that no gas is leaking from the tube. It will take several minutes to boil the water. When the water is boiling, the metal will be in thermal equilibrium with it, so record their temperature as $T_{m-o}$.

5. Record the ambient (room) temperature reading, from the second thermometer as $T_{a-o}$.

6. Fill the inner calorimeter can *about half–full* of *cold* tap water.

7. Measure the mass of the inner calorimeter can *with the water and stirrer in it*, and enter it in your data sheet.

8. Place the inner calorimeter, (with water and stirrer) into the insulating ring in the larger can (Figure 2). *Gently* put the second thermometer through the hole in the rubber stopper, and insert the stopper into the cover. Make sure the thermometer reaches near the bottom. Wait a few minutes until the inner calorimeter and the water are in thermal equilibrium; measure the temperature, and record it as $T_o$.

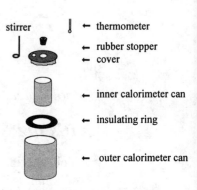

Figure 2

⊠  **Read Me First**
9.   When the water in the boiler has started to boil, let one team member lift the cover on the calorimeter assembly, while another person carefully lifts the string and **VERY  QUICKLY** transfers the masses into the inner calorimeter.  Then the first member **VERY  QUICKLY** covers it again.

10.   *S-l-o-w-l-y* stir with an up and down motion.  Watch the thermometer ; when the temperature *stops rising* , the system (calorimeter + water + metal) has reached thermal equilibrium.  Record this equilibrium temperature as $T_E$ .

11.   Again, record the ambient (room) temperature, using the first thermometer, as $T_{a-f}$

# Show-the-World:

1.   Calculate the mass of the water inside the calorimeter:

$$m_w = (m_{cal} + m_{stir} + m_w) - (m_{cal} + m_{stir})$$

2.   Calculate the  HEAT GAINED  by the water:

$$Q_w = m_w\, c_w\, (T_E - T_o)$$

3.   Calculate the  HEAT GAINED by the calorimeter:

$$Q_{cal} = m_{cal}\, c_{Al}\, (T_E - T_o)$$

4.   Calculate the  TOTAL  HEAT  GAINED  by the system (cal. + water):

TOTAL  HEAT GAINED

$$Q_{total} = Q_w + Q_{cal}$$

5.   If there have been no energy losses, then THERMAL ENERGY IS CONSERVED, i.e., the thermal energy lost by the metal must be equal to the total energy gained by the system of  (calorimeter + water).

*Conservation of Thermal Energy*

$$\text{HEAT LOST by metal} = \text{Total HEAT GAINED}$$
$$Q_{metal} = Q_{total}$$

6. Calculate the <u>specific heat</u> (c) of the <u>METAL</u> from the <u>HEAT</u> it <u>LOST</u>, by solving for $c_{metal}$ in:

$$\boxed{Q_{metal} = m_{metal}\ c_{metal}\ (T_{m-o} - T_E)}$$

7. Calculate the % error between your experimentally determined valued for the specific heat of the metal (for convenience call it $c_{exp}$), and the reference value (call it $c_{ref}$)

$$\% \text{ error} = \frac{C_{ref} - C_{exp}}{C_{ref}}\ 100\ \%$$

# DATA:          *Measuring Mother Nature*

$m_{cal}$  =  _____ gm.

$m_{stir}$  =  _____ gm.

$m_{(cal\ +\ stir)}$  =  _____ gm.

$m_{(cal\ +\ stir\ +\ water)}$  =  _____ gm.

$m_w$  =  _____ gm.

$m_{metal}$  =  _____ gm.       name of metal: _____

ambient temp. (at start) $T_{a-o}$  =  _____ °C

ambient temp. (at finish) $T_{a-f}$  =  _____ °C

original temp. of (water & cal.)    $T_o$  =  _____ °C

original temp. of metal    $T_{m-o}$  =  _____ °C

specific heat of water    $c_w$  =  _____    1.00    cal/gm °C

specific heat of Aluminum    $c_{Al}$  =  _____    0.22    cal/gm °C

Equilibrium temperature    $T_E$  =  _____ °C

heat GAINED by <u>water</u>                    $Q_w$    =    _____    cal.

heat GAINED by <u>calorimeter</u>              $Q_{cal}$    =    _____    cal.

<u>total</u> heat GAINED    $Q_{total}$    =    ($Q_w$    +    $Q_{cal}$)    =    _____    cal.


Conservation of Energy:          HEAT  LOST    =    HEAT  GAINED
                                 $Q_{metal}$    =    $Q_{total}$    =    _____    cal.


### *Specific Heat of Metal*

$c_{exp}$    =    _____    cal/gm °C

$c_{ref}$    =    _____ cal/gm °C              % error    =    _____    %


# QUESTIONS:        *Did you Learn Something ?*

1.   What do you think were important reasons that may have contributed to the error between your experimental value and the reference value for the specific heat of the metal ? Discuss.




2.  If you had four bars of metal, instead of two, would this extra mass have resulted in an *increase* or a *decrease* in your experimental value of specific heat ?   Explain why.

3.  If you had filled the inner calorimeter cup to the brim, would the effect have been an *increase* or a *decrease* in your experimental value of specific heat ?    Explain why.

4.  What was the purpose of asking you to measure the ambient temperature at the beginning and at the end of the experiment ?

5.  Why is water used as a coolant in a car's radiator ?

Did you ever notice that air feels a little warmer just after it rains or snows ? Have you ever wondered why ice-cubes can sit in your cold drink without melting ? It's not necessarily true that when you add heat to something its temperature has to rise ! If you add heat (thermal energy) to a solid at its melting point, *the temperature remains constant until it's completeley melted !*

For every gram of ice ($H_2O$) that toy want to melt (at 0°C), you would have to add about 80 calories of heat, but, for every gram of Lead (Pb) that you melt (at 327 °C), you only have to add about *6 calories* ! The basic reason why this difference is so large, has to ultimately be explained, like any good scientific theory, in terms their molecular structure.

Atoms of Lead are not holding each other as tightly together as are the molecules of ice. It doesn't take much heat energy to shake the Iron atoms apart (i.e., melt it) ; on the other hand, the ice molecules are electrically attracted to each other 14 times as much as the Lead, which means you need 14 times more calories, for each gram, to break them apart (melt) !

You're about to discover, for yourself, experimentally, exactly how many calories per gram (called the Latent Heat of Fusion) it takes to melt ice !

# The TOOL-BOX:

Aluminum Calorimeter Set:
(Outer Can, Inner Can,
Insulating Ring, Stirrer, Cover,
Rubber Stopper w. Hole)

2  0°–100° thermometers
  Bunsen Burner with Hose
  Stand, Clamp, Heating Pad
  Copper Boiler

2 Metal Rods
Mass Balance
String
Scissors

(Figure 1)

*tap water*                    *ice cubes*                    *equilibrium*

## What's the Recipe ?

✷ You can find the *Latent Heat of Fusion of water-ice* by throwing some ice-cubes into a cup of water. By measuring your masses, and your original and final (equilibrium) temperatures , the Heat ( $Q_w$ ) exchanged by the *water*, and the Heat lost by the calorimeter, can be calculated.

The heat (Q) transferred when the temperature changes is :

$$Q = m c \Delta T$$

where

$m \Rightarrow$ mass

$c \Rightarrow$ specific heat

$\Delta T \Rightarrow$ change in temperature

✳ By applying Conservation of Energy in the form of :

$$\boxed{\textbf{HEAT LOST } = \textbf{ HEAT GAINED}}$$

you can determine how much heat is needed to melt the ice.

✳ The Latent Heat of Fusion is just that amount of *heat per unit mass* :

$$L_F = \frac{Q_F}{m}$$

### The Ice Age

You can get a better sense of what's happening to the ice, by examining its evolution on a graph of *temperature* vs. *heat*, (Fig. 2):

*At its melting temperature* , the ice absorbs heat from the warmer water and calorimeter,

BUT ITS TEMPERATURE *DOESN'T CHANGE* !!!

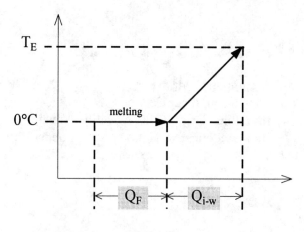

Figure 2

Only after it's *completely melted* does its temperature rise with added heat.

# STEP-BY-STEP:

1.   Measure, separately, and record in your data sheet, the *mass* of
     a)  the empty inner calorimeter can,
     b)  the stirrer.

2.   Read the  ambient (room) temperature, and record it.

3.   Fill the  inner calorimeter can *about half-full*  of *warm (not hot)* tap water ;
     measure its *mass* and *temperature*.  Record these in your data sheet.

4.   Place the inner calorimeter, (with water and  stirrer) into the insulating ring in the  larger
     can.  *Gently* put the thermometer through the hole in the rubber stopper, and insert
     the stopper into the cover.  Make sure the thermometer reaches near the bottom.

5.   Put *two to four*  ice-cubes on a paper towel, and wipe off the water on them with
     another paper towel.

6.   Add the ice-cubes to the inner calorimeter, and put it back into the calorimeter unit.
     Make sure you cover it.

7.   *S-l-o-w-l-y*  stir with an up and down motion.  Watch the thermometer ;
     when the temperature  *stops dropping* , the system (calorimeter + water + metal)
     has reached thermal equilibrium.  Record this equilibrium temperature as  $T_E$ .

8.   Remove the inner calorimeter, with the water and stirrer in it, and measure its combined mass.
     Record it as   $m_{tot}$  in your data sheet.

9.   Again, record the ambient (room) temperature, using the first thermometer, as  $T_{a-e}$

# Show-the-World:

1.   Calculate the mass of the *original* warm water inside the calorimeter:

$$m_{orig-water}  =  (m_{cal}  +  m_{stir}  +  m_w )  -  (m_{cal}  + m_{stir})$$

2.   Calculate the  HEAT LOST  by the original water:

$$Q_w  =  m_w \ c_w \ (T_o  -  T_E)$$

3.   Calculate the  HEAT LOST by the calorimeter and stirrer:

$$Q_{(cal + stir)}  =  m_{(cal + stir)} \ c_{Al} \ (T_o  -  T_E)$$

4.  Calculate the  <u>TOTAL</u>  HEAT <u>LOST</u> by the <u>system</u> (cal. + water + original water):

TOTAL  HEAT <u>LOST</u>

$$Q_{total} = Q_w + Q_{(cal + stir)}$$

)))))➡  <u>ASSUMPTION</u> : *We are going to assume that the temperature of the ice is at 0°C*

5.  If there have been no energy losses, then  THERMAL ENERGY IS CONSERVED !
    i.e., the thermal energy <u>lost</u>  by the system of (cal + stirrer + original water)
    must be equal to the thermal energy  <u>gained</u> .

Total <u>HEAT LOST</u> (by cal + stirrer + original water)  =  <u>HEAT GAINED</u>

$$Q_{(cal + stir + orig. water)} = Q_{GAINED}$$

<u>*GAINED BY WHAT*</u> ???

➡       a.   Part of the thermal energy first goes into breaking the molecular bonds that
             fuse the  $H_2O$ molecules together in a rigid crystal structure, **while the
             temperature remains constant !!!**        $(Q_F)$

➡       b.   The rest of the thermal energy goes toward heating the now-melted ice-
             water (at  0 °C) to the final equilibrium temperature.

6.  Calculate the  <u>mass</u> of the ice, by subtracting the mass of the (cal. + stir. + orig. water)  from
    the total mass :

$$m_{ice} = m_{tot} - m_{(cal. + stir. + orig. water)}$$

7.  Calculate the  HEAT  GAINED  by the  <u>melted  ice-water</u> :

$$Q_{i-w} = m_{ice} \ c_{ice} \ (T_E - 0.0 \ °C)$$

8.  Calculate  $Q_F$ , the  HEAT  required to completely melt the ice at the constant temperature
    of  0.0 °C

$$Q_F = Q_{GAINED} - Q_{i-w}$$

9.   Calculate $L_F$ , the Latent Heat of Fusion of ice-water :

$$L_F \;=\; \frac{Q_F}{m_i}$$

10.  Calculate the % error between your experimentally determined value for the Latent Heat of Fusion of Ice-Water  (for convenience call it  $L_{exp}$), and the reference value (call it  $L_{ref}$)

$$\% \text{ error} \;=\; \frac{L_{ref} \;-\; L_{exp}}{L_{ref}} \; 100\,\%$$

# DATA:        *Measuring Mother Nature*

$m_{cal}$  =  _____ gm.

$m_{stir}$  =  _____ gm.

$m_{(cal + stir)}$  =  _____ gm.

$m_{(cal + stir + original\ water)}$  =  _____ gm.

$m_{orig.\ water}$  =  _____ gm.

$m_{total}$  =  _____ gm.

$m_{ice}$  =  _____ gm.

ambient temp.  (at start)      $T_{a-o}$  =  _____ °C

ambient temp.  (at finish)     $T_{a-f}$  =  _____ °C

temp. of original water  &  cal.      $T_o$  =  _____ °C

original temp. of  ice      $T_{ice}$  =  _____ °C

specific heat of <u>water</u>                    $c_w$  =    1.00     cal/gm °C

specific heat of <u>Aluminum</u>                 $c_{Al}$  =    0.22     cal/gm °C

specific heat of <u>ice</u>                      $c_{ice}$  =    0.51     cal/gm °C

Equilibrium temperature        $T_E$  = _____ °C

     heat LOST by *original* <u>water</u>        $Q_w$  = _____ cal.

     heat LOST by <u>calorimeter</u>          $Q_{cal}$  = _____ cal.

     <u>total</u>  heat <u>LOST</u>        $Q_{LOST}$  =   ($Q_w$  +  $Q_{cal}$)  = _____ cal.

<u>Conservation of Energy</u>:        HEAT *GAINED*  =  HEAT *LOST*

                                    $Q_{GAINED}$  =   $Q_{LOST}$  = _____ cal.

heat GAINED by <u>ice-water</u>        $Q_{i\text{-}w}$  = _____ cal.

heat  ABSORBED to <u>melt ice</u> (at  T  =  0 °C):   $Q_F$  = _____ cal.

<u>EXPERIMENTAL</u>   Latent Heat of Fusion of ice-water  ($L_F$)

             $L_{EXP}$  = _____ cal/gm.

<u>REFERENCE</u>        Latent Heat of Fusion of ice-water  ($L_F$)

             $L_{REF}$  = _____ cal/gm.

     % error  = _____ %

# QUESTIONS:        *Did you Learn Something ?*

1.  What was the purpose of wiping the ice-cubes dry before putting them into the water ?

2.  What do you think were important reasons that may have contributed to the error between your experimental value and the reference value for the  LATENT HEAT of FUSION of ICE-WATER ?  Discuss.

3.  If you had filled the inner calorimeter with almost a full cup of tap water, instead of about a half-cup,  would this extra water have caused an increase or a decrease in your experimental value of  $L_F$ ?    Explain why.

4.  If our assumption about the temperature of the ice being at  0 °C  wasn't realistic, what effect would this have on your experimental  $L_F$ ?  For example, assume the ice temperature was  $T_{ice}$   =   −10 °C .

    a)  Would this have resulted in an *increase* or a decrease in your experimental value of  $L_F$ ?

    b)  Explain your answer.

    c)  Draw a set of axes.  Put "T" on the vertical axis, and "Q" on the horizontal axis. Sketch, using a set of connected straight lines, what happens as the temperature changes from  −10 °C  to  +10 °C .   Make sure your sketch is fully labeled.

# RESISTANCE and RESISTIVITY

**14**

If you've ever watched the flow of traffic as a road changes, suddenly, from asphalt to dirt, you probably noticed the flow slow, i.e., the number of cars per second moving past any point decreased. This is analogous to what happens when the same electric current (charge per second) moves through different materials of the same shape. The road's "roughness" is due to many factors (number & size of rocks, flatness, moisture, etc.). The electrical "roughness" of a material is called it's *resistivity* ($\rho$) , and is different for different materials because of the different atomic and molecular structures.

But that's not the end of the story. We can make two more analogies. If the road were four lanes wide, instead of two, the flow rate would be doubled ; the same would be true if the conducting path (the wire) were wider. Finally, traveling along a two-mile road will take you about twice as much time as it would if the same type of road were only one mile long. Does doubling the *length* of the electrical path double the time it takes for charges to travel that path ?

Notice that the last two characteristics, width and length, are geometrical ; it is just these additional qualities that distinguishes *resistance* from *resistivity* , as you're about to discover, experimentally.

# The TOOL-BOX:

Multimeter
Nichrome Wire mounted on Board

Various Leads
Unknown Resistor

## Part A   *Dial Your Ohms* - *(Multimeter as an Ohm-meter)*

There are several different methods you could use to measure electrical resistance. Two often used (but not necessarily more precise) methods are using a Voltmeter and an Ammeter to enable you to calculate the resistance. The Ohmmeter provides this current internally, usually through batteries. A single instrument that allows you to measure all three, for various ranges is called a multi-meter. Inside of it, you'd see a big rotary switch, to which are connected carious combinations of resistors corresponding to the various ranges, a Galvanometer, and batteries.

# On the face of it . . .

To use the multi-meter as an Ohmmeter, you just have to read one scale.  But notice that the

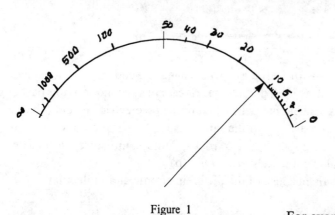

Figure 1

scale is not linear.  When the needle appears between two labeled divisions, divide the difference between those numbers, by the number of segments between them.  This will give you the value of each unlabeled division.  Then observe the largest division the needle has reached and just add its value to the smaller of the two numbered divisions.

For example,  this meter reads  12 , but that doesn't necessarily mean  12 Ω .

# 'Ohm on the Range'

You also have the ability to dial which range of resistance values is appropriate for the value you are measuring.  At the lower right of the dial selections, you'll find the ' Ω ' choices:

$$x 1 , \ x 10 , \ x 100 , \ x 1K , \ and \ x 10K$$

These are the multipliers.  The resistance, in *ohms* , is the value  you read from the scale, times the multiplier :

$$R = \textit{(scale reading)} \ x \ \textit{(multiplier)}$$

In our example, if the multiplier was set at  x 10 , then the resistance would read as :

$$R = ( 12 ) x ( 10 ) = 120 \, \Omega$$

# ✡ The Twilight Zone

*Q : But what if I'm measuring an unknown resistance, and  the needle doesn't move ?*

*A :*   If the resistance is in the *hundreds of ohms*, and you've selected the *x 10K* multiplier, the needle is reading only about  1/100  of it's full-scale !  The Ohmmeter is working, but it's difficult to read it.  To cure the problem, just switch to a *smaller multiplier*, such as  *x 100* , and the needle will swing over to the central part of the scale, making it easier for you to read.

## ✌ 𝒯𝓌𝓸 𝒻𝓸𝓇 𝓉𝒽𝑒 𝒫𝓇𝒾𝒸𝑒 𝓸𝒻 𝒪𝓃𝑒

In " *The Twilight Zone* " example above, had you switched to the *x 1K* multiplier, you would also be able to read the needle position, but the needle would have swung over to a different place.

*Q:*  *So which multiplier do I use*

A:  Theoretically, it shouldn't matter, but if you look closely, you'll notice that when you switch multipliers, the resistance you determine will be slightly different.  The reason is because switching the dial changes the electrical path used inside the meter.  Different paths involve different combinations of resistors, and therefore different degrees of error.  This is the "price-performance" factor that has to be considered when you buy and/or design any machine.

To minimize the error, choose the multiplier that causes the needle to rest *anywhere in the central three-quarters  (3/4)*  of the scale.

# STEP-BY-STEP:

## Part A   𝒟𝒾𝒶𝓁 𝒴𝓸𝓊𝓇 𝒪𝒽𝓂𝓈 - *(Multimeter as an Ohm-meter)*

Using the Ohmmeter, measure the resistance of the unknown resistor, using two different multipliers, and record your scale readings, multipliers, and R-values in your data sheet.

## Part B   How 𝓛𝒪𝒩𝒢 is your 𝑅 ?

Now that you know how to use the multi-meter as an Ohmmeter, let's use it to test what the effect of an object's *length* has on it's resistance.  The wooden board you have, also shown in Figure 2 , has mounted upon it, five metal posts, labeled "D" through "H" .

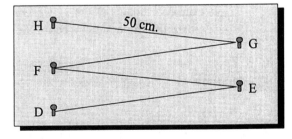

There is one continuous piece of Nichrome wire strung across them, such that the length of the wire between any two consecutive letters is  50 cm.

1.  Turn the dial to the  "Ohms (Ω)" section, and choose the *x1* multiplier.

*Figure 2*

2.  Touch the probes of the Ohmmeter to segment  D-E , and record it's resistance as  $R_{DE}$ , to the appropriate number of significant figures, in your data sheet.

3.  Keep repeating the procedure, for the following additional sets of points, in order to
    measure the resistance between them :
    >  points " D "   and   " F " .  Record this as  $R_{DF}$  in your data sheet.
    >  points " D "   and   " G " .  Record this as  $R_{DG}$  in your data sheet.
    >  points " D "   and   " H " .  Record this as  $R_{DH}$  in your data sheet.

    > Make sure you're reading the
    >    *correct scale   &*
    >    *scale  multiplying  factor*.

# Show the World !

## Part A    *The Multimeter*

Calculate the resistance of the unknown resistor, using the *scale reading* and *multiplier* you
used in each of your two different measurements, and enter them in your data sheet.

## Part B    *Resistance vs. Length*

1.  Calculate the circular cross-sectional *area* of the wire from :

$$ \text{Area}  =  \pi\, r^2  =  1/4\, \pi\, (\text{diam.})^2 $$

2.  Plot a graph showing Resistance (in ) on the vertical axis, and length (in cm.) on the
    horizontal axis.

3.  Draw, and measure the slope of, the straight line that best represents your data points.
    Display the coordinates of the *two points on your straight line* that you are using to
    calculate the slope with.  Record the slope on both your graph and data sheets.

4.  Multiply your measured slope by the *Area* you calculated in step #1 .

5.  Calculate the % difference  between your answer in the previous step, and  $100 \times 10^{-6}$ :

$$ \%\text{ difference}  =  \frac{100 \times 10^{-6}  -  R_{exp}}{100 \times 10^{-6}}  \quad 100\,\% $$

# DATA :   *Measuring Mother Nature*

## Part  A:   *Multimeter as an  Ohm-meter*

| scale reading | multiplier | R (Ω) |
|---|---|---|
|  |  |  |

## Part  B:   *Resistance  vs.  Length*

| segment | scale reading | multiplier | R (Ω) | length (cm.) |
|---|---|---|---|---|
| **D-E** |  |  | $R_{DE}$ = |  |
| **D-F** |  |  | $R_{DF}$ = |  |
| **D-G** |  |  | $R_{DG}$ = |  |
| **D-H** |  |  | $R_{DH}$ = |  |

diam.  =  _____ cm.

r   =  _____ cm.

A   =  _____ cm.$^2$

slope coord.    $L_1$  =  _____ cm.    $L_2$  =  _____ cm.

(from graph)

$R_1$  =  _____ Ω    $R_2$  =  _____ Ω

slope  =  _____

(Area)  x  (slope)  =  _____

% difference  =  _____ %

# QUESTIONS:          *Did you Learn Something ?*

1.  What are the *units* of the *slope* of your graph ?

2.  What are the *units* of  (Area) x (slope) ?

3.  What *physical quantity* do the units of your answer to Question # 2  represent ?

4.  Why do you think we chose Nichrome wire instead of Copper wire, for this experiment ?

5.  We assumed that the resistance of the posts and meter probes was zero.  Do you think that
    this was a  reasonable assumption to make ?  If so, why ?  If not, why not ?

6.  *Which segment* of Nichrome wire, of all the possible segments you tested, would cause
    the *largest current* to flow if a battery is connected to the board ?

It's taken millions of years for the human brain to evolve the complex state it's in now.  But for all it's complexity, our brains can only make sense of what we see when things:    a)  appear right-side up, instead of upside-down,  and
b)  respond to a stimulus in a simple linear way.

For example:  a)  the appearance of an upside-down face is barely recognizable. (Twist your head and neck, and look at your friend's face !)
b)  it would be difficult to estimate the temperature if the scales were logarithmic !  (Half the distance between two degree marks is not half of the temperature difference between them !)

For electrical devices, the two ideas translate to :
a)  *polarity* , i.e., does reversing the an applied voltage make the device respond differently ?
b)  *Ohm's Law*, i.e., does the resistance of a device remain constant over a range of applied voltages ?

You're about to discover the answers to these questions, experimentally, for two objects:
1)  the carbon resistor
2)  the semi-conductor diode

# The TOOL-BOX:

Low Voltage DC Power Supply

3 Resistors:    $R_1$  =  120 $\Omega$
( 5 % , 2 W)    $R_2$  =  240 $\Omega$
$R_3$  =  360 $\Omega$

Ammeter
DC Voltmeter
Various Leads
Switch
Semi-conductor Diode

# STEP-BY-STEP:

*Seeing  Colors!     (Counting  your  RRRRR's)*

Nobody would ever say the electrical industry is dull.  Besides being pretty, colors are used as a standard for identifying the resistance of simple Carbon resistors, such as the ones you're using.  The colors are in the form of circular rings around the cylindrical resistors.

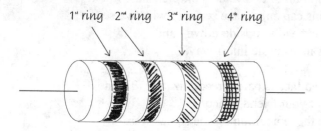

1ˢᵗ ring    2ⁿᵈ ring    3ʳᵈ ring    4ᵗʰ ring

Figure  1

Here's how to decode it:

1ˢᵗ  ring's color  ⟹  1ˢᵗ digit of the resistance

2ⁿᵈ  ring's color  ⟹  2ⁿᵈ digit of the resistance

3ʳᵈ  ring's color  ⟹  number of zero's

4ᵗʰ  ring's color  ⟹  tolerance of stated value

| ring  color | value | ring  color | value |
|:---:|:---:|:---:|:---:|
| Black | 0 | Green | 5 |
| Brown | 1 | Blue | 6 |
| Red | 2 | Violet | 7 |
| Orange | 3 | Gray | 8 |
| Yellow | 4 | White | 9 |

I wonder how much
tolerance I have ???

Tolerance

Silver       10 %

Gold          5 %

Ex 1:    A 10 %  tolerance means the *real* value of "R" falls within the range between

90 % and 110 % of the *stated* value of "R".

Ex 2:        Red    Blue    Yellow    Gold                Implies :

2        6        4        5 %            R  =  260,000 Ω  ± 13,000 Ω

247,000 Ω  ≤  R  ≤  273,000 Ω

Observe the color bands on each of the three resistors to make sure you are using the same values of resistance and tolerance as in the " Toolbox " list.  Record the colors in your data sheet in the same order you see them on the resistors.

# Part  A   *Polarity*

Some electrical devices, like a battery, have a "built-in" polarity, i.e., one terminal is at a higher potential than the other.  The higher potential terminal is marked  " + " .  The other terminal is marked as  " − " .  This doesn't literally mean that this terminal's voltage is negative, it's just there to remind us that whatever it's voltage is, it is *less* than the voltage of the other terminal.

Let's test two common electrical elements,  the *carbon resistor* , and the *semi-conductor diode* to see what the nature of their polarity is.  The test circuit we're going to use consists of:

a)   a *power supply*, to provide a source of *EMF* ,

b)   a *switch*,

c)   a *DC-ammeter*, to connect in  *series*  with the devices, in order to measure the current flowing through them.

d)   a *Voltmeter*, to connect in  *parallel*  to the devices, in order to measure the potential difference across them.

✓   You can use the **multimeter** as *either* an  ammeter  *or*  a  voltmeter, by turning the dial      to the appropriate section.  You will also have to choose the multiplier that allows you to easily read the needle's position on the scale that corresponds to that multiplier.

## ✓  *SAFETY  TIP :*

*Higher*

*is Better !*

> Always start a measurement with a *higher multiplier* that you think you'll need, and then switch to progressively lower multipliers until you can obtain a reading.  This prevents the needle from breaking or getting stuck when it swings all the way over if you've chosen too low a setting !

## *The  Carbon  Resistor*

Since the Carbon resistor is composed of just one material, Carbon, it doesn't have an intrinsic (built-in) polarity like a battery.  But when it's connected in a circuit, a *potential difference* develops across it, and it's only then that  " + " .  and  " − "  signs should be used to indicate this "imposed polarity.

A1.  Set up the circuit as shown in
        Figure 2 , but <u>do not turn the
        power on yet</u> ! Are you taking
        into consideration the polarity
        of each component ?  When you
        are ready, ask your friendly
        professor to check the wiring of
        your circuit.

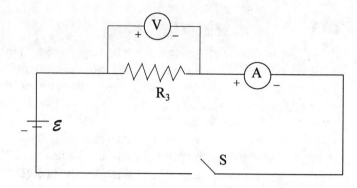

Figure 2

A2.  Turn the Power Supply on, and
        close the switch <u>briefly</u> ;  adjust
        the voltage on the power supply
        until you can read  1.0 volts on the voltmeter.  Observe the readings on both the Ammeter and
        the Voltmeter, carefully.  Make sure you are reading the correct scale and scale multiplying
        factors.  Record the values of current and voltage, to the appropriate number of significant
        figures, in your data sheet.  Open the switch.

A3.  Now  REVERSE  the resistor, and repeat steps  A1 and A2 .

## *The Semi-conductor Diode*

        The semi-conductor diode has the physical appearance of a resistor, but is not made of one
material.  It's actually made of two compounds in contact with each other ;  one is a piece of Germanium
with a "sprinkling" of Silicon atoms, and the other is a piece of Silicon with a "sprinkling" of
Germanium atoms.  When these pieces are put into contact with each other, an interesting effect occurs :
when a potential difference is put across it, the current flowing it is not the same as when the polarity is
reversed.  Let's experiment with it's polarity !

A4.  Replace the resistor in your test circuit,
        with the  semi-conductor diode, as
        shown in Figure 3 , but <u>do not turn
        the power on yet</u> !  When you are
        ready, ask your friendly Professor
        to check the wiring of your circuit.

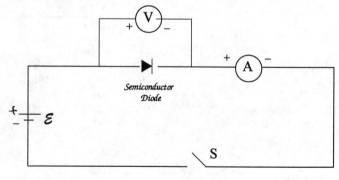

A5.  Repeat steps  A2 and A3 , but for the
        diode.

Figure  3

# Part B   *Ohm's Law*

> *"Sheriff Ohm says 'its the Law' "*

B1.   Remove the diode and replace it with the  240 Ω  resistor.  ($R_2$  =  240 Ω )

B2.   Repeat the same step  #2  that you performed when you had the first resistor in the circuit.

B3.   Close the switch, and change the setting on the Power Supply so that you read  2.0 volts on the Voltmeter.  Observe and record the values indicated on both the Ammeter and the Voltmeter.  Open the switch.

B4.   Keep repeating this procedure, but in each case  *increase*  the voltage across the resistor by 1.0 volt, until you have reached  10.0 volts.  Make sure the switch is  *open*  when you are not observing a meter reading.

# Show-the-World:
## Part B   *Ohm's Law*

1.   *Calculate*  the ratio:  $\dfrac{V}{I}$ for each pair of data values, and record them in your data sheet.

2.   *Plot*  an appropriately labeled and scaled graph of your ten sets of data.  Put  *Voltage* (V)  on the  *vertical axis*,  and  *Current* (I)  on the  *horizontal axis*.

3.   *Draw*, using a ruler,  *a straight line*  which best represents the average of all of your graphed data points.

4.   *Choose*  two points  *on*  your straight line, and record their coordinates in your data sheet.

5.   *Measure the slope*  of your straight line using your chosen two pair of coordinates, and record it in your data sheet as  $R_{exp}$ .

6.   *Calculate the  % difference*  between your slope, and the stated value of the resistor (let's call it the reference value  $R_{ref}$ .

$$\% \text{ difference} \quad = \quad \frac{R_{ref} \; - \; R_{exp}}{R_{ref}} \quad 100\,\%$$

# DATA:    *Measuring Mother Nature*

## COLOR of BAND

|  | 1st | 2nd | 3rd | 4th |
|---|---|---|---|---|
| $R_1$ |  |  |  |  |
| $R_2$ |  |  |  |  |
| $R_3$ |  |  |  |  |

### meaning :

|  | R ($\Omega$) | $\pm$ | $\Delta R$ ($\Omega$) |
|---|---|---|---|
| $R_1$ |  | $\pm$ |  |
| $R_2$ |  | $\pm$ |  |
| $R_3$ |  | $\pm$ |  |

## Part A  *Polarity*

### *The Carbon Resistor*

|  | V (volts) | I (Amps) |
|---|---|---|
| ORIGINAL position | 1.0 |  |
| REVERSED position | 1.0 |  |

### *The Semi-conductor Diode*

|  | V (volts) | I (Amps) |
|---|---|---|
| ORIGINAL position | 1.0 |  |
| REVERSED position | 1.0 |  |

## Part B  *Ohm's Law*        $R_{ref}$  =  $R_2$  =  240 Ω  (± 5 %)

| V (volts) | I (Amps) | $\dfrac{V}{I}$ |
|:---:|:---:|:---:|
| 1.0 | | |
| 2.0 | | |
| 3.0 | | |
| 4.0 | | |
| 5.0 | | |
| 6.0 | | |
| 7.0 | | |
| 8.0 | | |
| 9.0 | | |
| 10.0 | | |

_____
<u>slope coordinates</u>
(from graph):
_____

$I_1$ = _____ Amps          $I_2$ = _____ Amps

$V_1$ = _____ Volts          $V_2$ = _____ Volts

slope  =  _____

% difference  =  _____

# QUESTIONS:          *Did you Learn Something ?*

1.  What would happen if you kept the switch closed for a long time (such as 10 minutes) ?

2.  Compare the resistance of the *resistor* in the reversed position versus the original position.

3.  Compare the resistance of the *semi-conductor diode* in the reversed position versus the original position.

4.  Do your 10 data points support Ohm's Law ? Explain why, or why not.

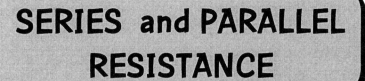

# SERIES and PARALLEL RESISTANCE

## 16

Open up your computer, or your radio, or your TV, or your digital watch, and you'll find more than one resistor. Why would you need more than one ? All that a single resistor can do is limit the amount of current flowing through it, but when resistors are combined with other resistors, or with capacitors, or types of circuit elements you are soon to learn about, they play different roles in the circuit. For example: your radio and TV receives many different frequencies of electrical vibrations from the antenna, but it takes only a few resistors and capacitors to filter out all of them except the one you've tuned to ! Now, if you had a network of $10^{10}$ resistors and capacitors, you would have a terrific communication system, wouldn't you ? Well *you do* ! It's your nervous system, and its CPU (Central Processing Unit) is your brain ! Each neuron (cell) has a resistance, with an electro-chemical potential difference between the inside and the outside of it.

To understand what effects are produced by connecting large numbers of resistors together, we need to look at the two basic ways of connecting them: *series*, and *parallel*. You're going to explore, experimentally, what happens when you combine resistors in a simple series circuit, a simple parallel circuit, and then a combination of the two.

*The YIN-YANG of Resistance*

# The TOOL-BOX:

| | | |
|---|---|---|
| 6 V Battery | | Ammeter |
| 3 Resistors: | $R_1$ = 120 $\Omega$ | Voltmeter |
| $\left(\begin{array}{c}5\ \%\ ,\ 2\ W\\ \text{each}\end{array}\right)$ | $R_2$ = 240 $\Omega$ | Various Leads |
| | $R_3$ = 360 $\Omega$ | Switch |

# STEP-by-STEP:

## *Seeing Colors !* *(Counting your RRRRR's)*

Observe the color bands on each of the three resistors to make sure you are using the same values of resistance and tolerance as in the Tool-Box list. Record the colors in your data sheet in the same order you see them on the resistors.

# Part A  SERIES RESISTANCE

## CIRCUIT

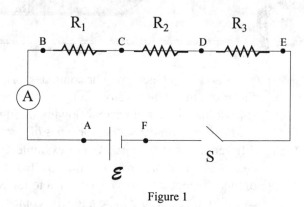

Figure 1

1.  Set up the series circuit as shown in Figure 1 , but <u>do not close the switch yet</u>.  Are you taking into consideration the polarity of each component ?  When you are ready, ask your friendly Professor to check the wiring of your circuit.

2.  Close the switch, and record the reading on the Ammeter.  (Make sure you are reading the correct scale and scale multiplying factor.  Record the current, to the appropriate number of significant figures, in your data sheet.

3.  Take the probe tips of the Voltmeter in your hands, and simply touch the following sets of points, in order to measure the potential difference between them:
    a) points " B " and " C " .  Record this as  $V_{BC}$  in your data sheet.
    b) points " C " and " D " .  Record this as  $V_{CD}$  in your data sheet.
    c) points " D " and " E " .  Record this as  $V_{DE}$  in your data sheet.
    d) points " F " and " A " .  Record this as  $V_{FA}$  in your data sheet.

# Part B  PARALLEL RESISTANCE CIRCUIT

1.  Set up the parallel circuit as shown in Figure 2 , but <u>do not close the switch yet</u>.  Are you taking into consideration the polarity of each component ?  When you are ready, ask your friendly Professor to check the wiring of your circuit.

2.  Close the switch, and record the reading on  the Ammeter.  (Make sure you are reading the correct scale and scale multiplying factor.  Record the current, to the appropriate number of significant figures, in your data sheet.

3.  Take the probe tips of the Voltmeter in your hands, and simply touch the following of points, in order to measure the potential difference between them:

    a) points " C " and " D " .  Record this as  $V_{CD}$  in your data sheet.
    b) points " B " and " E " .  Record this as  $V_{BE}$  in your data sheet.
    c) points " A " and " F " .  Record this as  $V_{AF}$  in your data sheet.

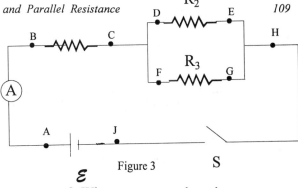

Figure 3

# Part C   *Compound*

## RESISTANCE CIRCUIT

1. Set up the compound resistance circuit as shown in Figure 3 , but <u>do not close the switch yet</u>. Are you taking into consideration the polarity of each component ?  When you are ready, ask your friendly Professor to check the wiring of your circuit.

2. Close the switch, and record the reading on  the Ammeter. (Make sure you are reading the correct scale and scale multiplying factor.  Record the current, to the appropriate number of significant figures, in your data sheet.

3. Take the probe tips of the Voltmeter in your hands, and simply touch the following of points, in order to measure the potential difference between them:
   a) points " B " and " C " .  Record this as  $V_{BC}$  in your data sheet.
   b) points " D " and " E " .  Record this as  $V_{DE}$  in your data sheet.
   c) points " F " and " G " .  Record this as  $V_{FG}$  in your data sheet.
   d) points " C " and " H " .  Record this as  $V_{CH}$  in your data sheet.
   E) points " B " and " H " .  Record this as  $V_{BH}$  in your data sheet.
   F) points " A " and " J " .  Record this as  $V_{AJ}$  in your data sheet.

# Show-the-World:

## Part A          *Resistors  in  SERIES*

1. Calculate    $\dfrac{V_{AF}}{I}$    , and record it in your data sheet as  $R_{exp}$ .

2. Calculate the  *equivalent resistance of the entire circuit* , using two different formulas, and record the answers in your data sheet as  " $R_{th}$ "    $\Rightarrow$   theoretical resistance.

   FORMULA  # 1 :

   $$R  =  R_1  +  R_2  +  R_3$$

   FORMULA  # 2 :

   $$\frac{1}{R}  =  \frac{1}{R_1}  +  \frac{1}{R_2}  +  \frac{1}{R_3}$$

3. a)  Calculate the  % difference  between your answer in calculation # 1  and your answer   using formula # 1 , and enter the result in your data sheet.

   b)  Calculate the  % difference  between your answer in calculation # 1  and your answer   using formula # 2 , and enter the result in your data sheet.

$$\% \text{ difference} = \frac{R_{ref} - R_{exp}}{R_{ref}} \, 100\,\%$$

# Part B          *Resistors in PARALLEL*

Repeat, and record the same four types of calculations you made in *Part A)*

# Part C          *Resistors in a COMPOUND Circuit*

Repeat, and record the same four types of calculations you made in *Part A)*

# DATA:      *Measuring Mother Nature*

# Part A          *Resistors in SERIES*

$\mathcal{E}$ = _____ Volts          I = _____ Amps

$V_{BC}$ = _____ Volts

$V_{CD}$ = _____ Volts

$V_{DE}$ = _____ Volts

$V_{FA}$ = _____ Volts          $R = \dfrac{V_{FA}}{I}$ = _____ Ω

FORMULA # 1 :   $R_{th}$ = _____ Ω          % difference = _____ %

FORMULA # 2 :   $R_{th}$ = _____ Ω          % difference = _____ %

## Part B          *Resistors in PARALLEL*

$\mathcal{E}$ = _____ Volts                    I = _____ Amps

$V_{CD}$ = _____ Volts

$V_{BD}$ = _____ Volts

$V_{AF}$ = _____ Volts                    $R = \dfrac{V_{FA}}{I}$ = _____ $\Omega$

FORMULA # 1 :   $R_{th}$ = _____ $\Omega$              % difference = _____ %

FORMULA # 2 :   $R_{th}$ = _____ $\Omega$              % difference = _____ %

_____                    _____

## Part C          *Resistors in a COMPOUND Circuit*

$\mathcal{E}$ = _____ Volts                    I = _____ Amps

$V_{BC}$ = _____ Volts

$V_{CD}$ = _____ Volts

$V_{FG}$ = _____ Volts

$V_{CH}$ = _____ Volts

$V_{BH}$ = _____ Volts                    $R = \dfrac{V_{AJ}}{I}$ = _____ $\Omega$

$V_{AJ}$ = _____ Volts

FORMULA # 1 :   $R_{th}$ = _____ $\Omega$              % difference = _____ %

FORMULA # 2 :   $R_{th}$ = _____ $\Omega$              % difference = _____ %

## Part A  *Polarity - The Carbon Resistor*

|  | 1st | 2nd | 3rd | 4th |
|---|---|---|---|---|
| R₁ |  |  |  |  |
| R₂ |  |  |  |  |

|  | $V$ (volts) | $I$ (Amps) |
|---|---|---|
| ORIGINAL position | 1.0 |  |
| REVERSED position | 1.0 |  |

## Part B  *Ohm's Law*        $R_{ref}$ = $R_2$ = 240 Ω ( ± 5 % )

| $V$ (volts) | $I$ (Amps) | $\dfrac{V}{I}$ |
|---|---|---|
| 1.0 |  |  |
| 2.0 |  |  |

# QUESTIONS:        *Did you Learn Something ?*

1.  In **Part A**,   $V_{BC}$ represents the voltage across _____ .

    $V_{CD}$ represents the voltage across _____ .

    $V_{DE}$ represents the voltage across _____ .

    $V_{FA}$ represents the voltage across _____ .

In **Part B**,         $V_{CD}$ represents the voltage across         _____  .

                       $V_{BE}$ represents the voltage across         _____  .

                       $V_{AF}$ represents the voltage across         _____  .


In **Part C**,         $V_{BC}$ represents the voltage across         _____  .

                       $V_{DE}$ represents the voltage across         _____  .

                       $V_{FG}$ represents the voltage across         _____  .

                       $V_{CH}$ represents the voltage across         _____  .

                       $V_{BH}$ represents the voltage across         _____  .

                       $V_{AJ}$ represents the voltage across         _____  .


2.   Was there a difference between the stated *EMF* of the battery, and the *terminal voltage* that you measured ?  If yes, how do you explain it ?


3.   ( <u>MULTIPLE CHOICE</u>:  Choose either  a , b , or  c )
       When resistors are connected in parallel, the equivalent resistance is :
           a.   greater than the largest value of resistance.
           b.   less than the smallest value of resistance.
           c.   between the smallest and the largest values of resistance.

       Explain why your answer must be true.

4.  For each part of the experiment, write out the formula which produced the *smallest* % difference in the chart below :

| Type of circuit | Formula | % difference |
|---|---|---|
| Part A:   SERIES | | |
| Part B:   PARALLEL | | |
| Part C:   COMPOUND | | |

5.  In **Part C** ,

   (a)    Which components of the compound circuit are connected in parallel ?

   (b)    Identify those voltages, of the six voltages you measured, that are equal to within

        a   10 %   difference.

## 17

# LAW of REFRACTION

Have you ever tried to catch a fish with a net, and just missed ? It's not that the fish was faster than you, but you probably didn't put your net in deep enough ! The real depth was greater than the apparent depth. Why ? What makes rainbows ? What causes mirages like the "*liquid road*" on a hot day ? Why do stars "*twinkle*" ? Why do cameras, telescopes, and microscopes all have lenses inside them ? The explanation of all of these phenomena is due to one single fact : LIGHT BENDS WHEN IT ENTERS A DIFFERENT MATERIAL ! The reason light bends (REFRACTS), is because it's SPEED CHANGES, even though it's frequency remains the same.

We can compare the light's speed in air to it's speed in various materials, as a way of indexing (cataloging) how much it refracts with respect to a dotted line drawn perpendicular to the boundary between the two materials. In other words, the index of refraction ( n = c / v ) of each material, together with the angle of incidence (measured from the normal) determines what the angle of refraction (measured from the normal) will be. You're going to discover this relationship experimentally for a piece of glass.

# The TOOL-BOX:

Flat Rectangular Plate of Glass     Ruler     2 Sheets of Blank Paper
4 Pins     Protractor     1 Sheet of Cardboard
0.5 mW, Neon LASER     Sharp Pencil     Semi-circular Plastic Tray

# STEP-BY-STEP:

## Part A *Index of Refraction of Glass*

1. Put a sheet of paper on top of the cardboard, and place the glass plate near the center of the paper.

2. Carefully outline the top and bottom edges of the glass plate with your pencil.

3.  Place a pin vertically into the paper, roughly  3 - 4 cms. from the upper-left side of the glass plate, as in Figure 1 .  Let's call it *pin A*.

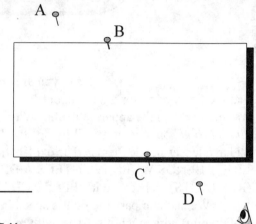

4.  Place another pin vertically into the paper *just at the edge* of the upper side of the glass plate, but displaced to the right by about  2 cms.  Let's call it *pin B*.

  *AVOID the " WHOOPS ! "*

**BE CAREFUL NOT TO MOVE THE GLASS !!**

Figure 1

5.  Now, bend down so you can see the image of the two pins *from the front side* of the glass plate.  Your eyes should be at the same vertical level as the glass.  Move your head sideways until you can line up the images of pins  A  and  B  in your sight.  With your partner's help, place a third pin (call it  pin C) vertically into the paper *just at the edge* of the glass *but also in the same line of sight* as the images of pins  A  and  B .

# DON'T GET <sup>UP</sup> YET !

6.  Finally, place a fourth pin  (call it pin  D) vertically into the paper about  3 - 4 cms. in front of the near side of the glass *but at a point where pin  D  lines up with the images of pins  A  and  B* (see Figure 1  again).

## Let's Draw !

7.  Remove the glass plate and the four pins, and draw the following straight lines with your ruler and pencil :
    a.  Draw a straight line from  point A  to  point B .
    b.  Draw a straight line from  point B  to  point C .
    c.  Draw a straight line from  point C  to  point D .

## SEEING DOTS IS NORMAL !

8.  Line up the base of your protractor with the higher horizontal line which represents the edge of the glass.  Center the protractor at  *point B* , and make a light dot with the pencil at a point corresponding to the  90°  position.  Using the ruler, draw a  *dotted*  line from the pencil dot *straight through* point B , *until you reach the bottom horizontal line.*  This is the  <u>*normal line*</u> .

9.  Rotate the sheet until  point C  is at the
    top, and repeat step # 8 in order to
    draw another dotted normal line
    that passes through  point C .  Look
    at Figure 2 .

10. Using the ruler
    a)  draw extensions of lines A-B
        and  C-D  *using solid lines*, so
        that each has a length of
        3.5 inches,

    b)  extend line A-B with a  4-inch
        *dotted line* on the right side of
        point B ,

    c)  draw a solid line *perpendicular*
        to line  C-D  *but only between
        line  C-D  and  the dotted line you just drew from Part (b)* .  Label this line  "d" .  It's called
        the  "*parallel displacement*"  because it shows you how much the light ray,  emerging *from
        the glass back into the air*, has shifted sideways from it's original direction when it first
        *entered the glass from the air*.

Figure 2

11. Label the angle between line  A-B  and the normal line through point B  as  $\theta_1$ .
    Label the angle between line  B-C  and the normal line through point B  as  $\theta_2$ .
    Label the angle between line  B-C  and the normal line through point C  as  $\theta_3$ .
    Label the angle between line  B-C  and the normal line through point C  as  $\theta_4$ .

## *Measuring Up* !
12. Using the ruler, measure the length of  "d"  and enter it into your data sheet.

13. Using the protractor, *carefully*  measure the four angles:      $\theta_1$ , $\theta_2$ , $\theta_3$ , and $\theta_4$ , and record them
    in your data sheet.

# Part B:    ✽★☆ *Total Internal Reflection* ☆★✽

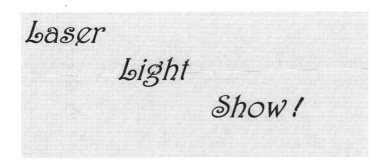

L.A.S.E.R.
**L**ight
**A**mplification
    by the
**S**timulated
**E**mission
    of
**R**adiation

B1.   Put water into the semi-circular tray until it's
        about 3/4 full.

B2.   Lay a book down on the lab table, thick enough to reach the level of the laser
        opening, and stack on top of it the following (in the same order) :
                    — the sheet of cardboard
                    — a blank sheet of paper
                    — the protractor
                    — the tray of water.

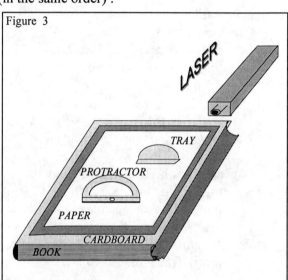

Figure 3

   Place the tray on top of the protractor.  Adjust the
   tray so that:
            a)  it's flat side lies along the <u>base-line</u>
                 (*not the bottom edge*) of the
                 protractor, and center the flat side
            b)  the center of the flat side is next to the
                 protractors center hole.  Place a pin
        at the center of the hole, for easy reference.

B3.   *Delicately*  place the LASER in the orientation
        shown in Figure 3 , and plut it into a
        <u>grounded</u> (three prong) outlet, <u>BUT **DO NOT** TURN IT ON YET</u> !!!

**✗✗✗   WARNING** !!!
**YOU MAY GO  <u>BLIND</u>   OR**
**HAVE OTHER  <u>SERIOUS EYE DAMAGE</u>**

**IF YOU <u>LOOK</u> *DIRECTLY* INTO**
**THE BEAM OF A LASER ,**
*EVEN IF THE BEAM*
*GOES THROUGH WATER FIRST* !!!!!!

B4.   When you are ready, have your friendly Professor check to see of everything is OK, before he/she
        gives permission to turn the LASER on.

B5.   Looking from above the water, slide the switch at the back of the LASER to the  ON position.  You
        should be able to see the red laser beam inside the water ;  if you can't, just *blow on the water*,
        or *mix it with your finger !*
   Rotate the *protractor* until you can see the laser beam hit the center of the flat side of the tray ;  the
        tray will rotate with it.

B6.  Now, have your partner *very slowly* rotate the *protractor  counter-clockwise*, 3° at a time,
ALWAYS making sure the laser beam hits close to the center (where the pin is) !
Observe the *refracted* ray, which passes from the water into the air.  The ray is invisible in air, but
you can just hold a sheet of paper perpendicular to the table to see the red spot where the
refracted ray strikes it !

Based on what you observe, is the angle of refraction
a)  *greater than* ,
b)  *less than*
c)  *equal*

to the angle of incidence ?

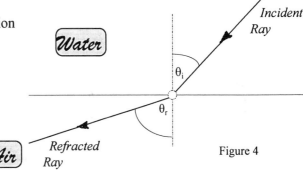

Figure 4

*Where'd it go ???*
*( **Not** seeing is believing ! )*

B7.  Continue *s-l-o-w-l-y* rotating , *UNTIL, amazingly, the refracted laser beam* **DISAPPEARS** *!!!*
As soon as this happens, tell your partner to <u>stop
rotating</u> .  The angle of incidence at which this
happens is called the *CRITICAL ANGLE* ;  it lies
between the incident ray, and the normal line, as
shown in Figure 5 .
Record it as  $\theta_C$  in your data sheet.
Notice that the angle of refraction is  90° .

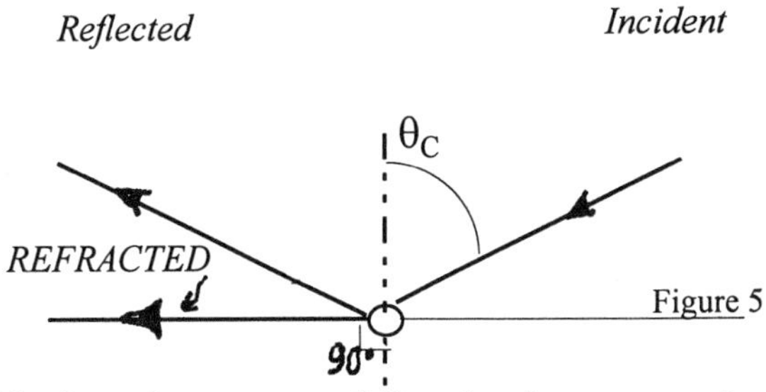

Figure 5

B8.  Have your partner rotate the protractor *back* a few
degrees, and then *s-l-o-w-l-y* forward again, but this time observe *carefully*, what happens to the
*reflected* ray, i.e., the ray that reflects off the flat side of the tray *back into the water*, just as the
angle of incidence becomes greater than the critical angle.  Describe what you observe in your
data sheet.

B9.  *One Good Turn Deserves Another !*

Continue to *s-l-o-w-l-y* rotate the protractor,  in 3° increments, for another  30° .  Watch carefully,
and describe what you observe.

# Part C:     *Reflections in the Water !*

C1. ***HOLD IT !***  Don't put the equipment away just yet !  Let's do one more experiment.  Let's
investigate how the angle of *reflection* varies with the angle of *incidence*.
Observe, and record the angle of *reflection* for the following five angles of incidence:
60° , 65° , 70° , 75° , and 80° .

C2. ***OK ! DONE !***  Slide the switch at the back of the LASER to the  OFF  position.

# Show-the-World:

## Part A:     *Index of Refraction*

A1.  Calculate:     $\sin \theta_1$ , $\sin \theta_2$ , $\sin \theta_3$ and $\sin \theta_4$ , and record them.

A2.  Calculate the experimental *index* of refraction of the glass by using the Law of Refraction :

$$n_{Air} \sin \theta_{Air} = n_G \sin \theta_G$$

Use     $n_{Air} = 1.00$     and     $\theta_{Air} = \theta_1$     as the angle of incidence.

A3.  Calculate the   % error   between your experimental value of  $n_G$  (call it  $n_{G\text{-}exp}$ )  and the reference
value of the index of refraction for this glass  (call it  $n_{G\text{-}ref}$ ) , which will be given to you by your
friendly Professor.

$$\% \text{ error} = \frac{n_{G\text{-}ref} - n_{G\text{-}exp}}{n_{G\text{-}ref}} \ 100 \ \%$$

## Part B:     *Total Internal Reflection*

B1.  Calculate  $\sin \theta_C$ ,   and record it.

B2.  Calculate your experimental *index* of refraction  ( $n_{G\text{-}exp}$ ) for your water, by applying the Law of
Refraction using :

$$n_{Air} = 1.00 \ , \qquad \sin \theta_C \ , \qquad \text{and} \qquad \theta_r = 90°$$

B3.  Calculate and record the    % difference   of your experimental value with respect to the reference
value for water :

$$n_{w\text{-}ref} \quad = \quad 1.33$$

# Part  C:   *Law  of  Reflection*

Calculate the    per-cent (%)  difference of your measured value of the angle of *reflection* , with respect
to the given angle of incidence, in each of the five cases, and record them in the data table.

# DATA:        *Measuring Mother Nature*

# Part  A:   *Index  of  Refraction  of  Glass*

$$d \quad = \quad \underline{\hspace{3cm}} \quad cm.$$

| $\theta$ (degrees) | $\sin \theta$ |
|---|---|
| $\theta_1 \quad =$ | $\sin \theta_1 \quad =$ |
| $\theta_2 \quad =$ | $\sin \theta_2 \quad =$ |
| $\theta_3 \quad =$ | $\sin \theta_3 \quad =$ |
| $\theta_4 \quad =$ | $\sin \theta_4 \quad =$ |

$$n_{G\text{-}exp} \quad = \quad \underline{\hspace{3cm}} \qquad n_{G\text{-}ref} \quad = \quad \underline{\hspace{3cm}}$$

$$\% \text{ error} \quad = \quad \underline{\hspace{3cm}} \quad \%$$

# Part  B:   *Total  Internal  Reflection*

$$\theta_C \quad = \quad \underline{\hspace{3cm}} \,^{\circ} \qquad \sin \theta_C \quad = \quad \underline{\hspace{3cm}}$$

$$n_{w\text{-}ref} \quad = \quad \underline{\quad 1.33 \quad} \qquad n_{w\text{-}exp} \quad = \quad \underline{\hspace{3cm}}$$

$$\% \text{ error} \quad = \quad \underline{\hspace{3cm}} \quad \%$$

## Part C:     *Law of Reflection*

| angle of incidence $\theta_{in}$ | 60.0° | 65.0° | 70.0° | 75.0° | 80.0° |
|---|---|---|---|---|---|
| angle of reflection $\theta$ | | | | | |

| % difference | | | | | |
|---|---|---|---|---|---|

# QUESTIONS:          *Did you Learn Something ?*

1.  Based on your data, and within experimental error, which angle was the same as
    a)  angle  $\theta_1$ ?
    b)  angle  $\theta_2$ ?

2.  If  $\theta_1$  was a larger angle than the one you used in your experiment, what effect would this have on " d " ?  Explain your answer.

3.  If you did the experiment with glass that had a higher index of refraction than the one in your experiment, what effect would this have on " d " ?  Explain your answer.

# Part  B:    *Total  Internal  Reflection*

What do you think happened to the refracted ray when it disappeared ?

What do you think happened when you exceeded the critical angle ?

Use your experimental value of the index of refraction  $(n_{w\text{-}exp})$  for the water, together with the known
    speed of light in air  ( $c$  =  $3.0 \times 10^8$ m/s ) , to calculate the speed of the LASER light in the water.

The *wavelength* of the  red LASER light  *in air*  is  $\lambda_{air}$  =  $6.328 \times 10^{-7}$ meters.  Use the reference
    values of the index of refraction of water and air, to calculate the *wavelength* of red LASER light
    when it's in the water.

# Part  C:    *Law  of  Reflection*

Does your data support the Law of Reflection, to within  5 %  ?
        Give reasons to defend your answer.

# Do You Know How to
# Draw a (*Physics*) – Graph ?

---

"A picture is worth a thousand words" . –  Its also worth a thousand numbers (or more) !
Do an experiment;  When you put the measured values of  x  and  y  in a data table, it might look like this:

| t (sec) | 1.0 | 2.0 | 3.0 | 4.0 | 5.0 | 6.0 | 7.0 |
|---------|-----|-----|-----|-----|------|------|------|
| y (cm)  | 2.3 | 4.4 | 7.1 | 9.4 | 11.2 | 14.0 | 16.6 |

If we know from theory that  " t "  and  " y "  are related, then Mother Nature gives us the right to display these values on a two-dimensional graph.

Stud:   Why bother ?  I already have the details !  besides, I have to answer the questions that analyze my results.

Prof:   *That's it !!*  You've just answered your own question !  Analyzing implies looking at the _trend_ in the relationship between  t  and  y  for _all_ of the  ( t , y )  pairs, not just one pair.

Stud:   But I can look at the trend by comparing each set of numbers in the data table.
Prof:   Would you be able to do this so easily if you had  20  or  30  pairs of numbers ?  Or  100  pairs ? Or  1000  pairs ?  What if  5  consecutive values of  y  increased, then the next two decreased, then the next  3  y-values increased, etc.   What would you make of that ?

Stud:   Hmm ... mm ..m.   ☹

Prof:   Don't blame yourself.  The limitation is not yours, its a human limitation characteristic of the *"logical-half"* of our brain.  But Mother Nature also gave us an *"artistic-half"* to our brain, and it is this part that gives us powerful insight into understanding natural phenomena.  We take advantage of it by drawing a picture, i.e., a graphical display from our numbers.

Stud:   OK,  OK,  I'm persuaded. Here's my graph:

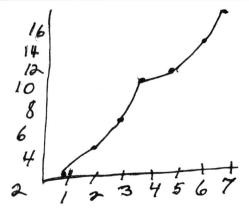

Prof:    You know what's wrong with this graph ?

Stud:   No, what ?

Prof:    EVERYTHING !!

Stud:   Everything ??  But my friends draw it the same way !  I also saw a similar graph in a business chart.

Prof:    How long did it take you to draw it ?
Stud:   About two minutes,  freehand.
          If everything is wrong, then what's the right way ?

Prof:    Keep in mind, that by drawing a graph we're not just making it easier to analyze the data for ourselves, but also for anyone else in the world that's interested !


Prof:    **LET'S COMPARE THE TWO GRAPHS.**

①    " Using Graphic Language "

Use graph paper for graphs !
for you to plot  and read values
are *precisely* separated from each

That's what it was designed for.  It's easier
more precisely, because adjacent grid lines
other by the same distance .


②    "........ through **thick** and `thin` "

The axes need to be drawn with a *sharp* pencil or pen      aligned *precisely* with a horizontal grid line and a vertical grid line, i.e., ***use a ruler*** !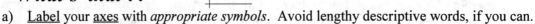


③    *" What's that ?? "*

a)  Label your axes with *appropriate symbols*.  Avoid lengthy descriptive words, if you can.

b)  Put the corresponding *units* , in parentheses, after the symbols.

c)  Title the entire graph. Make it brief and relevant. Use large lettering, and place it   near the top of your graph page, so it easily noticed.

     Ex:

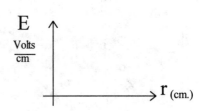

**Electric Field** vs. **radius**
in a
Solid Cylindrical Insulator

④  **"*Where were you born ?*"**
Label your origin with a zero !

0

⑤  **"*No one notices me !*"**
Put a *NEAT* little circle around each data point.

⑥  **DON'T  *"CONNECT  THE  DOTS"* !!!**

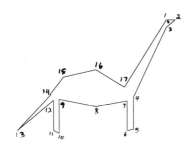

That's OK  for pictures in children's books or stock-market trends, but not here.  The numbers you measure in an experiment are an approximation to points on a CONTINUOUS THEORETICAL CURVE . What we need, then, is to draw the best continuous curve that approximates those points EVEN IF NONE OF THE DATA POINTS LIE ON THE APPROXIMATE CURVE WE DRAW *!!!*

If theoretically we expect a straight line, then we should draw an average straight line, with a ruler.

Stud:  OK , but exactly how ?

Prof:  Visualize each data point as having a *rubber-band* attached between it and the line we're going to draw.  With all of the rubber bands pulling on the line,  what angle and position would the line settle in at so that there would be no net movement of the line ?

It *can't* look like this, points on *one side*, would pull it toward

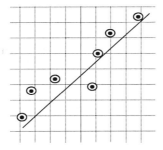

because with most of the data our imaginary rubber-bands that side (the left in this graph).

A fairer *representation* of the data points would look like this:

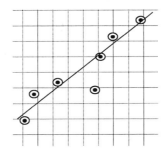

If theoretically we expect a curve, like a parabola or any curve that's not a straight line, then sketching an average curve is more difficult, but NOT HOPELESS !  A curved piece of plastic called a *French-Curve* helps a lot.  Just lay *a part* of *one* of the curves on the French-Curve that most closely resembles an average group of data points.  When the plastic curves too far away from om the the next group of data points you work on, just rotate and/or use a different curved segment.  You can get a very nice curve with just a little practice !

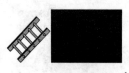

# " Scaling  the  Wall "

Here's a situation where a student performed an experiment measuring the position of an object " x " as a function of the elapsed time.

| t (sec) | 0.0 | 1.0 | 2.0 | 3.0 | 4.0 |
|---------|-----|-----|-----|-----|-----|
| x (cm)  | 0.0 | 1.5 | 2.3 | 2.8 | 3.4 |

# " Perfection  or  Defection" ?

Here is the student's graph:

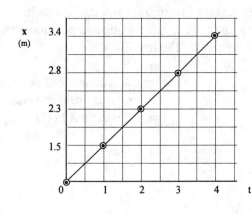

*Of course* it looks perfect ! ***Too perfect !!!***  As you advanced by one second (one square's length on the  horizontal  " t "- axis),  you incorrectly arranged the labeling so that you would advance by *exactly one box* on the vertical " x "- axis, in order to get to the next observed data mark.

Instead:

1.  Set up your axes with  *equally*

                                         *labeled*

                                              *intervals*

covering the <u>entire</u> range of values

you obtained experimentally.

2.  Locate as precisely as you can where your data values lie on your axes,

i.e.,        let your data values fall where they may !!

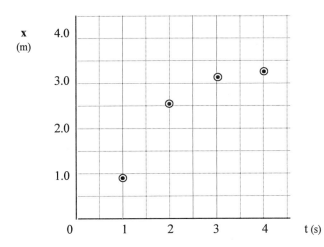

Looks  a lot different, doesn't it ?

Now you can see that the curve is
definitely  *not linear* !!

# The Slippery SLOPE

The Slippery SLOPE

How steep is it ?  If , as you ski down a hill, you descend by  3 feet  for every  2 feet  you move

 forward, then a simple way to describe the steepness of the hill is to say

it's  SLOPE  is:   $\dfrac{3 \text{ feet down}}{2 \text{ feet forward}}$

or,  using the conventional coordinate system:    y (+)↑ →x (+)    { $\Delta y$ **down** $=$ $- \Delta y$ **up** }

SLOPE  $=$   $\dfrac{3 \text{ ft. } (down)}{2 \text{ ft. (forward)}}$  $=$  $\dfrac{-1.5 \text{ ft.}}{1.0 \text{ ft.}}$  $=$  $-1.5$

To sum it up:

a)  For every  1.0 foot you  **RUN**  (forward) , you  **DROP**  (negatively rise)

1.5  feet .

b)  The  steepness of this hill is:    SLOPE  $=$  $-1.5$

In general:

**SLOPE**  $=$  $\dfrac{\text{RISE}}{\text{RUN}}$

where the  "**RISE**" can be a

*positive* (up) displacement

*or* a

*negative* ("down") displacement.

# STEP-BY-STEP:

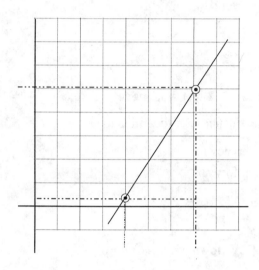

①   Pick two points *on* the line.

⇨   The best points to pick are those that are *closest* to the intersection of a horizontal and vertical line of the graph paper.

②   From each of the two points, draw *neat dashed* lines to intercept each axis perpendicularly (at 90°).

③   Extend the horizontal dashed line through the first point until it intersects the vertical dashed line of the second point.

④   To be more specific, let's imagine the graph to be a straight line of electric voltage (volts) vs. electric current (Amps).

Here are the data values obtained from an experiment. For clarity, let's put the *independent* variable (Voltage) on the horizontal axis, and the dependent variable (Current) on the vertical axis.

| V<br>(volts) | 1.0 | 2.0 | 3.0 | 4.0 | 5.0 | 6.0 |
|---|---|---|---|---|---|---|
| I<br>(Amps) | 4.8 | 16 | 34 | 43 | 53 | 68 |

Our goal is to measure the slope of the graph we draw.

## *Current  vs.  Voltage*

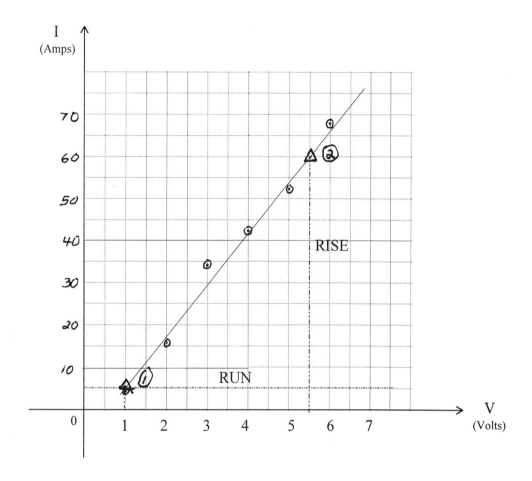

Two points *on the line* to use to determine slope are:

point ❶     $V_1$ = 1.0 V  ,    $I_1$ = 5.0 A

point ❷     $V_2$ = 5.5 V  ,    $I_2$ = 61 A

RUN  =  $\Delta V$  =  $V_2 - V_1$  =  5.5 V  –  1.0 V  =  4.5 V

RISE  =  $\Delta I$  =  $I_2 - I_1$  =  61 A  –  5.0 A  =  46 A

SLOPE  =  $\dfrac{\text{RISE}}{\text{RUN}}$  =  $\dfrac{46 \text{ A}}{4.5 \text{ V}}$  =  $10.2 \frac{A}{V}$